Bees Make *the* Best Pets

BEES MAKE *the* BEST PETS

All the Buzz about Being Resilient, Collaborative, Industrious, Generous, and Sweet—Straight from the Hive

JACK MINGO

Conari Press

Published by Conari Press, a division of Mango Media Inc.

Cover Design: Jim Warner
Cover Photograph: © Photosearch.com
Layout & Design: Dutton & Sherman

For permission requests, please contact the publisher at:
Mango Publishing Group
2850 S Douglas Road, 2nd Floor
Coral Gables, FL 33134 USA
info@mango.bz

For special orders, quantity sales, course adoptions and corporate sales, please email the publisher at sales@mango.bz. For trade and wholesale sales, please contact Ingram Publisher Services at customer.service@ingramcontent.com or +1.800.509.4887.

Bees Make the Best Pets: All the Buzz about Being Resilient, Collaborative, Industrious, Generous, and Sweet—Straight from the Hive

Library of Congress Cataloging-in-Publication Data is available on request.
ISBN: (print) 978-1-57324-625-5
BISAC category code: NAT017000, NATURE / Animals / Insects & Spiders

Printed in the United States of America

Dedicated to Reverend Lorenzo Lorraine Langstroth (1810–1895), whose revolutionary hive helped make modern beekeeping possible.

Prologue

Silent Night, Buzz Buzz Buzz Buzz

It's midnight on Christmas Eve and I'm outside in the cold getting ready to press my ear against a beehive. The reason may not make a lot of sense to you.

I'm not sure it makes a lot of sense to me, either. It's the result of a promise I'd made to myself in September, an easy commitment in daylight and warm weather. Tonight, though, I left a warm bed to stumble into the feral part of my backyard, the ground made treacherously uneven from a multi-year conspiracy of chickens and moles attacking from above and below. Worse, I am forced to depend on light from a street lamp and the neighbors' reflected Christmas displays because I haven't carried a light. There were two reasons for this. One, I hadn't wanted my neighbors to see what I was up to, and two, I didn't want to rouse the targets of my midnight reconnaissance: the bees.

Why am I doing this, you ask?

Okay, I'll come clean. I had read that long ago English beekeepers decked their hives with boughs of holly on Christmas

Eve. Then, late at night, they'd slink quietly back into the bee yard and wait. They did that, they said, because the bees serenaded them each year by quietly humming Christmas carols at the stroke of midnight.

It's not that I believed it, of course, but the story touched me deeply in ways I didn't fully understand. I *wanted* it to be true, in the same way I wanted everything to be true about Christmas. There's so much about the holiday that demands a total suspension of logic, from Santa to the half-god baby born to a virgin. Why not toss carol-singing bees into the mix?

To be honest, there was never a moment that I believed the bees would hum carols, but I liked the idea of the ceremony, the tradition of honoring insect friends by being with them at a special time on a significant night. I admit also that I was curious about how a story like that got started. Is it possible that random beehive sounds "upon a midnight clear" could really somehow sound like a glorious song of old? You know, like how your brain tries to hear words and patterns from radio static or crowd noises? Perhaps random buzzing, filtered through a mix of holiday spirituality and strong spirits, could sound something like "Adeste Fideles." I wanted to find that out, too.

I suspected that maybe the story was just a beekeepers' hoax, a way to reserve a quiet place away from others to think, drink, or meditate upon the season, but whatever the case, I intended to find out for myself tonight.

I stop a couple of yards away from the hives and notice how quiet everything is. On summer nights, you can hear the buzz of the hive at any hour, like a miniature factory running 24/7, as the bees ceaselessly clean, form wax combs, tend the babies, cool the hive, and dry the flower nectar into thick honey. On December 24, though, because the bees have no reason to cool the hive (on the contrary!) and no nectar to dry, the hives sound more like the ghost factories of Detroit, not just silent but freakishly so.

I step forward and squat down next to the hive. Still no sound. It occurs to me that old English beekeepers didn't use wooden boxes like mine but skeps woven from straw and reeds. Maybe sound carried better through them. Concerned about disturbing the bees, I hadn't intended to touch the hive at all, much less with my unprotected ear, but clearly I had no choice. It's almost midnight on Christmas Eve, and I don't want to have to wait another year to try again. So I gently brush my fingertips across the top of the hive, clearing off some of the dew and dirt, and gingerly lower my head, ear first, to the lid of the hive.

The surface is cold and damp and I'm a little afraid, but that's quickly forgotten because I hear something through the lid: the bees!

On December 24 the hives sound more like the ghost factories of Detroit, not just silent but freakishly so.

It's a different sound from the daytime hive, lower and more uniform, a steady, pulsing drone, like the sound of a . . . what? It's familiar and soothing, but I can't figure out what it sounds like. It *sort of* reminds me of the oddly comforting hum of a cheap 1970s-era electric clock that I used to have next to my bed, but that isn't quite it.

Still, I'm here to answer a question: do the bees make any sound that could be interpreted as a Christmas carol? I listen; the sound drones on in a pulsing monotone, never changing rhythm or pitch, the bees flexing their wing muscles to generate heat. This doesn't sound anything like carols. I reluctantly have to admit that Catholic monks singing a Gregorian chant—even Buddhist monks droning "Om"—are more musical than a beehive on a Christmas midnight. Even if nursing a half gallon of mead, there's no way an honest beekeeper could claim to hear a Christmas carol from that.

I sigh, frankly surprised that I'm disappointed. It's cold. I'm tired. My quest for knowing is over. I should go back inside. Yet, I stay and listen to the sound. There's something in that buzz that's deeper and older than Christmas carols, or even Christmas. The roar of a river? Not exactly. The roar of a lion? Closer, but more soothing.

I reluctantly have to admit that Catholic monks singing a Gregorian chant—even Buddhist monks droning "Om"—are more musical than a beehive on a Christmas midnight.

Then it strikes me: a beehive on a cold winter night, settled in for warmth, sounds like a purring cat. I suddenly realize that this purr existed long before house cats, or even before humanity was there to hear it. There are fossils of honey bees that are 23 to 56 million years old. Saber tooth tigers and mastodons may have heard this sound. In fact, some scientists believe bees may go even further back, all the way to the dinosaur era. I realize that maybe even sweet-toothed sugi-yamasauruses heard this sound.

The thought comes to me: it came upon a midnight clear, and it *is* a glorious song of old. In fact, this may be the oldest living sound I'll ever hear. So who needs carols?

The Key of Bee Natural

Adult bees, when they're inside the hive, make the sound of 190 vibrations per second, or a note halfway between the F# and G below middle C on the piano. That's not so interesting. What is fascinating, though, is this: when they fly, the tone bees make is—as it should be—B (248 vibrations per second).

> If bees fly in B natural, what note do they sting in? Bee sharp. What note when they hit the windshield? Bee flat.

More Bee Sounds (Last Word, I Promise)

Oddly, winter is the time when a beehive is most in tune. Most of the bees that winter over are fully grown female workers bunched together for warmth. During that time relatively few new bees are hatched. In the warmer parts of the year, a hive is made up of not just adult females, but also male drones, young females, and bees of all ages doing different jobs; each of those jobs create different sounds. Newly hatched females are full-sized, but their wings do not become fully hardened into flight-worthy tools until the age of nine days. When they fan their floppy new wings for warmth and ventilation, the lack of wind resistance means their wings fan faster than the adults' wings, making a higher tone. Meanwhile, the oversized drones have bigger wings that flap more slowly, creating a lower tone. The guard bees, protecting the hive from bears and beekeepers, fly fast in a beeline buzz bomb, in order to have the most impact when they give a warning thump and then a sting; this creates a higher, more insistent tone. Perhaps the time to imagine you can hear Christmas carols is in the summer, when there are more notes to choose from.

THE MOST PREVALENT NOTES

Very young bee fanning: C#–D
Adult guard bee attacking: C–C#
Adult bee flying: B
6-day old bee fanning: A–A#
Adult bee fanning: F#–G
Drone flying (loud, like a bronx cheer): Discordant flat low G

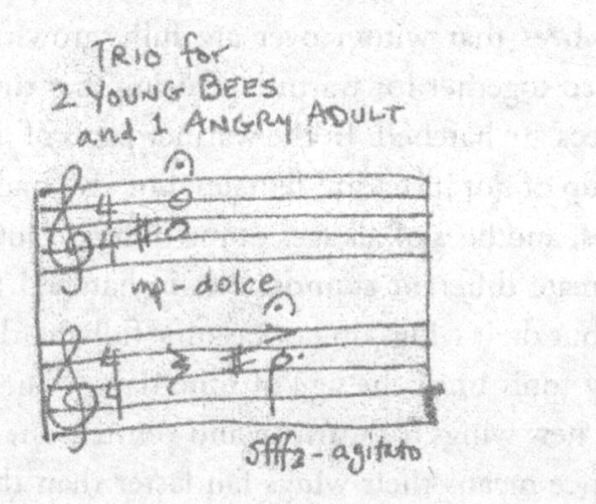

Keeps the Elephants Away

Why keep bees? Because it keeps the elephants away. Okay, it's an old joke, but just in case you haven't heard it:

> A man in a restaurant is perturbed by the odd behavior of a woman at the next table. He asks: "Why do you keep snapping your fingers and tossing your napkin in the air?"
>
> She answers: "Because it keeps the elephants away."
>
> "But that's ridiculous," says the man. "There's not a wild elephant for thousands of miles!"
>
> She gives him a triumphant look, and responds: "See how well it works?"

It turns out that bees really *can* keep the elephants away. In 2011, the BBC reported that Kenya had successfully reversed a serious decline in the elephant population, bringing their numbers up to 7,500. The problem was that pillaging pachyderms began raiding subsistence farmers' fields for tomatoes, potatoes, and corn. The destruction, as you can imagine, of having a bull elephant in your garden can be pretty extreme, and farmers began fighting back with guns and poisons.

The elephants easily knocked down fences and barriers, but in 2009 researchers at the University of Oxford and Save the Elephants discovered a method that was 97 percent effective in repelling elephants: beehives. A group of 17 farms was surrounded by a border of 170 beehives, placed 10 meters (33 feet) apart.

Elephants may be thick-skinned, but they don't like bees for a very good reason. The bees are very good at targeting the vulnerable parts of even thick-skinned animals, around the eyes, mouth, and nose. Elephant trunks are especially sensitive, and the aggressive African bees will fly right up inside them to sting if necessary.

The bottom line is that elephants attempted 32 raids over a three-year period. Only one got through the beeline; the rest were quickly convinced to pack up their trunks and go.

Last news was that conservationists intended to use the idea in other communities as well. Meanwhile, the farmers were trained to harvest honey and wax from their garden guardians, providing additional income and extra motivation for keeping the hives in good shape.

My Conversion

I don't often have road-to-Damascus, struck-by-lighting, instant-enlightenment, come-to-Jesus, scales-falling-from-the-eyes moments. However, my conversion to the Church of the Living Bees was one of those moments. (Say hallelujah, somebody!)

Friends, I had always been apathetic to bees at my strongest moments, slightly scared of them at my weakest. That all changed on a visit to an environmental awareness house in Berkeley, California in the summer of 1978, when I was a young man. It was there that I saw an observation hive.

Previously, I'd seen a few of these beehives with Plexiglas on the sides so you can see inside but had always been indifferent to them.

I loved ant farms, despite their mournful quality, because you could actually see the ants doing something recognizably antlike: digging freeform tunnels, eating, drinking, and carrying the bodies of their constantly expiring farm mates for burial. But beehives? There was no concrete activity I could really figure out. It just looked like modestly repugnant bugs running around randomly, like a lot of cockroaches scurrying around in a box.

Little did I know. I was a fool. Once lost, now found. Blind, but now can see.

What did it take? Somebody with a little knowledge and about two minutes. He pointed out a soccer ball-sized ring of yellows, oranges, and reds in the beeswax cells. This was the pollen that the bees had collected. The bright ring was a line of demarcation. Outside of it, the bees filled the comb cells with nectar; inside it was the nursery, containing eggs and larvae.

The outside of the circle was suddenly understandable. Field bees arrived from outside, carrying either nectar or pollen. Just inside, they transferred their cargo to warehouse workers who put it where it belonged. It was no less (and, frankly, no more) interesting than watching the loading docks of a warehouse.

The inside of the pollen circle, however, made up for it. The placement of the pollen is convenient because pollen is

Observation hive

what the nursemaid bees feed bee larvae—little white grub-like things, each inside its own cell. (Adult bees eat only honey.)

Then I saw the queen. She was surrounded by a small circle of worker bees. Each was facing her, looking like petals on a daisy. The queen was sticking her long, pointy ass into an empty cell, laying an egg. I waited and when she finished, she was pushed and pulled by her attendants to the next empty cell. I looked closely at the cell she had just left and, sure enough, there it was, the egg, white and smaller than a grain of fine sand.

I watched the nursery for quite a while. It was abuzz with activity: Eggs being laid, larvae squirming, and the bigger ones were being sealed into cells with a pollen/wax mix, to make the final transition to full-fledged bee.

It was so cool. I wanted my own observation hive. It would be a few years, but I eventually would get one of my own. But more about this later.

WHY BEES MAKE THE BEST PETS, TAKE 1

1. They are among the nicest stinging insects you'll ever meet.
2. A bee is (literally) as cute as a bug. Actually cuter than most.
3. Honey.
4. Beeswax.

continued

5. They protect your garden from elephants. Seriously. They may work with other animals, too.
6. They have soft fuzz that is almost irresistible. It glistens in the sun for great photos.
7. They are amazing natural architects.
8. They make their own building materials.
9. They are ten times cleaner than any pet you've ever owned.
10. They make your house smell better instead of worse.

Bee Team #1: The Sex Workers

There are only two categories of bees that have the same job their entire life, and for both of them, that job is reproduction.

The Queen: The queen lays eggs. She's the only one that does, and she does it nearly constantly during honey season—from 1,000 to 1,500 eggs a day. She's not really the leader of the hive; more like its ovaries. Still, she's very important to the continuing existence of the hive so she is well protected, pretty much to the point of house arrest. When she's doing well, she communicates it through cues of scent and behavior; if she's not doing well, that news gets quickly disseminated through the hive as well. The workers show who's really boss at that time: They immediately begin planning to depose her, creating special, extra large queen cells that look like peanuts sticking out from the other cells. They grab a healthy-looking larva for each and drench it in an extra dollop of royal jelly that makes it grow faster, bigger, and with fully developed sex organs. The first queen that emerges kills off her rivals, and goes on a mating flight to meet some cute drones and make sweet but fatal love to them. She comes back with enough sperm to last a lifetime of egg-laying.

Now this part is kind of cool: The sperm stays in a special repository from which she can *at will* decide whether to fertilize an egg or not. Why the choice? Weirdly, it's for gender selection. If she chooses to fertilize an egg, it hatches a female worker bee. If she chooses to not fertilize an egg, it hatches a drone.

The queen lays eggs. She's the only one that does, and she does it nearly constantly during honey season.

BLUE BLOODS

The queen isn't the only blue blood in a beehive. It turns out that bee blood really is a greenish-blue color. So I make a point to honor each of them as royalty, from the lowliest nursery worker to soaring superstar foragers.

Drones: Drones are the few males in a hive, and they play up the role like the pampered gigolos they are, hanging around, doing no work, living off the work of their sisters. Each hive in an area provides drones that head to a designated drone area, waiting for any virgin queen to fly by looking for a good time. The variety of drones hanging around somewhat minimizes the chance of in-breeding with their queenly sisters, but these things happen even in the best of hives. Still, there are some interesting varieties of sexual experience among bees that you won't read about in the Kama Sutra. Mating in midair, for example, swooping and diving toward the ground. Unfortunately, the drones' pleasure is even more short-lived than most males. As they withdraw, they discover that their penises are still inextricably stuck inside the queen. When they tear away from their lover, they *really* tear away from her, crumpling to the ground in a painful death. Any drones that manage to survive into the fall are given the boot to die in the cold as winter approaches. Cruel, but understandable.

Bee Team #2: Hive Got a Job for You

Not every bee can hang around laying eggs or anticipating sex. In fact, worker bees normally don't have the opportunity to do either one. The workers, all female, live a life of celibacy and constant work, literally from the moment they're born until the moment they're thrown out of the hive. It's almost an extreme prototype for the modern corporation.

But it's not as if there are no promotion possibilities in the HiveCo Corporation. Every worker bee goes through a gamut of jobs in its short life:

Custodian: As soon as a worker bee eats her way out of her brood cell, she has a job: cleaning out the cell she just came out of, making it ready for the next egg.

Nursemaid: Next job is taking care of the grubby little babies, the larvae: feeding them, keeping them the right temperature, dosing them with royal jelly, and sealing them up into their cells with a mix of wax and pollen.

Warehouse worker: Receiving, moving around, and storing the honey and pollen is a full-time job.

Heating, cooling, and ventilation: Bees flapping their wings as they work creates a constant stream of fresh air circulating through the hives. On hot days, lines of bees wind throughout the hive, all facing the same direction, holding tight and buzzing their wings as if flying, creating a small wind that blows in through the entrance, around the hive, and out again. Some bees hold small drops of water to cool the air further. On cold days, bees crowd close in the nursery, flexing their wing muscles to generate heat, keeping themselves, the queen, and the babies warm.

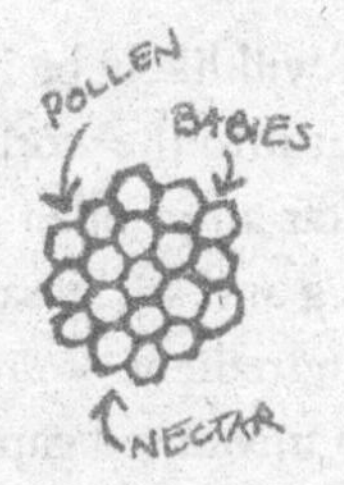

TACTICAL PROCEDURE, BEE VS. PREDATOR

DEFCON 4: Wary monitoring from hive.
DEFCON 3: Menacing buzzes around intruder's head.
DEFCON 2: Repeated head-butting impacts against intruder's face and head.
DEFCON 1: Commencement of stinging.

Guards: Standing just inside the entrances of a hive, the guards act like bouncers in a club, keeping out unauthorized bugs. When threatened by something bigger, they fearlessly fly forth to defend the hive from birds and animals of all kinds, including people, bears, elephants, and pretty much anything else that gets too close.

Builders: By the twelfth day of its life, a bee's wax glands begin working, dropping little fish-scale-looking flakes of wax all over the hive, like oversized dandruff. The builders collect the flakes, crawl into unfinished honeycomb cells and shape the wax into cell walls.

Field bees: In what will likely be its last job, the fully mature bee begins going out into the world to collect needed supplies. Most notably nectar and pollen from flowers, of course, but also water for cooling and sticky sap from plants and trees. The latter is the major ingredient of *propolis*, a sticky filler material bees use to seal up unwanted gaps and holes.

AGING BEES LOSE THEIR HAIR TOO

Here's how to tell a honey bee's age:

1. Where did you find it? If outside the hive, it's at least nine days old. If it's in the hive but outside the ring of pollen, it's at least four or five days old. If it's inside the circle, it's probably a nursemaid, which is usually only a few days old, unless it's the big one, the laying queen. She can be anywhere from a few weeks to several years old.
2. If the bee is extremely fuzzy, it's young. If its hair is thin enough that its black body is showing through the down, it's old—probably a month or more. In bee years, during the hard-working summertime, a month is the equivalent of about fifty to sixty human years.

There Are Always Exceptions

- Remember that I said worker bees don't "normally" lay eggs? That's because the queen's pheromones suppress the reproductive systems of the workers. However, if the queen dies and there are no larvae that can be groomed to replace her, that can change. Unless a beekeeper intervenes with a new queen, the hive is doomed. In that case, a terminally queenless colony will try to spread its genes before it dwindles away, using an unexpected tactic: some of the workers will start laying eggs. However, since they haven't mated, their unfertilized eggs will yield only drones (as mentioned above). Maybe some of them will get lucky and find a willing queen, passing the hive's genes along in its dying days.

- When older bees begin collecting nectar and pollen from outside the hive, their brains change, and not really for the better. For example, after they memorize the surroundings of the hive, they lose the ability to learn new things. Normally, they stay that way until they die. However, sometimes "normal" gets disrupted; for example, if a hive has to grow a new queen, there can be a month-long gap before any new bees hatch. Normally, that would mean the larvae from the new queen wouldn't have young nursery workers available to take care of them, and they'd die. In that case, some of the field bees return to the nursery worker job. Here's where it gets interesting: researchers from Arizona State University discovered that going back to larvae-rearing makes their old brains work again like young brains, restoring their mental agility and ability to learn. (Interestingly, many human grandparents have discovered that keeping up with their fleet-footed, quick-witted grandkids has a similar effect.)

THE ROYAL SENDOFF

When a new queen is ready to fly off and mate, there's an interesting pageantry in the beehive that looks reminiscent of royal weddings. Suddenly thousands of worker bees pour out of the hive at once. Those that can fly take to the air, flying excitedly in circles above the hive; the young flightless bees congregate on and around the landing board like well-wishers. They are not aggressive, and you could almost interpret their behavior as celebratory (and who knows, it might be). However, this grouping and flying has survived as a behavior because it serves a purpose. It's dangerous to leave the hive, especially for a big, juicy, slow-moving queen, because many birds love eating bees. By coming out in large numbers, there's a variety of tasty looking targets, so the queen has a much better chance of surviving, even if a hundred bees get eaten during that time. If the queen is only one of 10,000 or more crawling and flying around, the odds are only one in a hundred that she'll be one of the unlucky ones.

continued

The King Is Dead, Long Live the Queen

"If there's a queen bee, is there a king bee?" People have had strange ideas about bees for eons. In the first century AD, the poet Virgil wrote a guide to beekeeping in which he stated that bees reproduced asexually. This idea that honey and beeswax came from a sexually chaste source, mirroring their belief about the mother of Jesus, was particularly appealing to the Catholic Church. As a result, beehives were commonly kept around missions and convents, and only beeswax candles were considered unadulterated enough to burn on church altars.

Beehives back then were kept in skeps, one-piece woven baskets covered with hardened mud or dung, which didn't allow much investigation into the inner workings of hive life. At some point somebody noticed that there was a bigger bee of some kind in there. The assumption, based on the society of the time, was that it must be the king that ruled over the hive and told the workers what to do. It wasn't until the 1500s that it was suggested that the king bee was not only female, but also the one that laid all the eggs. Still, the assumption didn't change that she was a virgin queen, ruling over the hive population of male workers. Consider the surprise of male scientists in the 1670s, messing around with a new invention called a microscope, when one of them discovered that all of the workers were females.

Still, thanks to the "queen" label, most people assume that the queen bee is the leader of the hive. Not true. The hive is essentially a well functioning anarchist community, with bees doing what needs to be done based on their biology and instincts. Decisions are not made by dictate or consensus; they're hardwired into the bees. When something needs to be done, one or more of them does it. The queen is an important member of the hive with a singular job to do, as well as the mother of all the workers, but she is as much the prisoner of the hive as its queen. And her mates, the drones, aren't even princes—they're more like disposable consorts. Very disposable, it turns out.

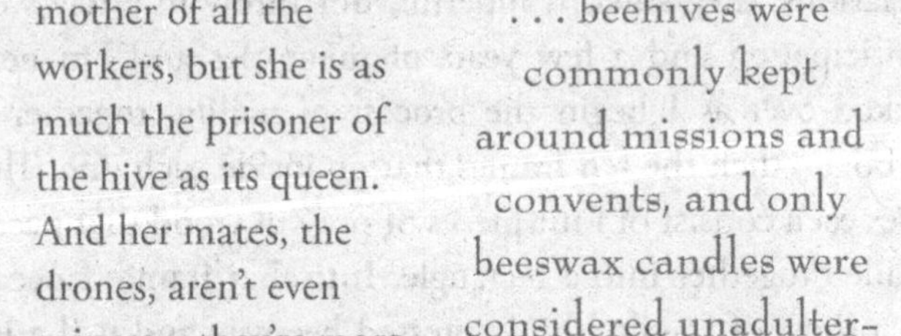

> . . . beehives were commonly kept around missions and convents, and only beeswax candles were considered unadulterated enough to burn on church altars.

Deconstructing Bee Construction

Who knows for whom the doorbell tolls? Actually, a brown-uniformed guy with heavy boxes knows: it tolls for me.

The boxes on my front porch have come from a beekeeping supply house, Dadant, the country's biggest and oldest. As I move them into the house, they make the sounds of wood rattling together like the top notes of a marimba. I'm expanding my bee yard, adding another hive, and adding some honey supers to the others. But first there's a lot of pounding and painting to do.

Opening the cardboard boxes, I'm greeted by the sweet smells of pinewood and fresh beeswax. Even devoid of mental associations, they smell wonderful, but throw in a few weeks of anticipation and a few years of memories and I'm nearly knocked over as I begin the process of nailing together the hive boxes, then the ten frames that go inside each one. Those frames each consist of four pieces of pre-cut wood that need to be nailed together into a rectangle. Into that frame, I need to coax a sheet of fragile, thinly pressed beeswax and nail a long narrow strip that theoretically holds the wax in place. Most of the job is not hard, but it is repetitious. It is hard to mess up the frame too badly, but it is easy to wreck the wax sheet. Barely thicker than construction paper, it can crack by accidentally flexing it a little too far while laying it into the frame, or clumsily punching a hole through it with the hammer while pounding in the tiny nails, poking out just a millimeter or two from the wax, that barely hold the thin strip of wood that barely holds the wax in place, hopefully long enough that the bees will cement it into place before it falls out.

First Beehive

When I mention my beekeeping, people sometimes ask "how did you get started?" (and sometimes "In the name of God, why?!?"). I wish I had a story that shows that I'd had noble environmental intentions and coherent motivation. But honestly, I got into beekeeping in the same way I've made a lot of big decisions of life, love, career, and philosophy—by stumbling into them in an unfocused, whimsical, and alarmingly superficial manner.

When I think about it, most of those decisions turned out all right. Either there's method in my madness, or I've been pretty damned lucky. As a case study, here's how I stumbled into beekeeping.

Ant Farm on Steroids

Blame the shortcomings of ant farms for my introduction to beekeeping.

Remember ant farms? My several siblings and I got one as a present one year. At first it was fascinating, elbowing the others to get a better vantage point while watching the ants digging new tunnels through the virgin sand. Not much time passed, though, before the farm collectively became just another ignored and neglected pet. The tipping point came when the soil was thoroughly dug and the farm became a depressing existential hell filled with depressed and moping ants with absolutely no purpose in the world. Worse, their numbers slowly dwindled, as they died off one by one to be buried by the survivors—for reasons not fully explained—in the northeastern corner of the farm. When the farm got down to just one of the social insects listlessly pushing a clump of dirt pointlessly from point A to point B, it was more than a tender heart could bear.

The ideal would've been to have gotten a queen ant. They lay eggs so the colony wouldn't die out. They also have pheromones that keep an ant farm both motivated and populated. But ant suppliers don't provide queens and when digging up my own ants I never found a queen, so I finally gave up on ant farms.

Then, a decade after my last ant farm, I found something even better, something like an ant farm on steroids: a fully functioning observation beehive. The best part of it was that these bees looked happy and industrious, and I soon saw the reason: they had a queen. They had eggs and larvae. It was a working, thriving community that looked as if it were purposeful instead of tragic. I wanted one.

I Got One . . . Sort Of

It was only a year or two later that I saw a newspaper feature about a local guy who built and sold observation hives. I called him up right away, drove to his house in San Francisco, and bought one. Well, the box anyway. No bees. He suggested finding a local beekeeper and asking to buy a couple of frames of bees and a queen. No problem.

With some effort, I found the name of a beekeeper. I had already arranged to keep the bees in the library of a small counterculture "hippie" school where I taught middle and high school students art, phys ed, history, film, and so on for a very modest wage. One snag: when I called him, he wouldn't sell me just a couple of frames. What he said made sense: "An observation hive is so small that it will be just barely self-sustaining. You'll probably need to add bees to it now and again, trade out old frames and add new ones full of larvae if you lose your queen, and so on." He suggested I get a whole, standard beehive and use it to keep the observation hive going. "You might even get a little honey out of it," he added.

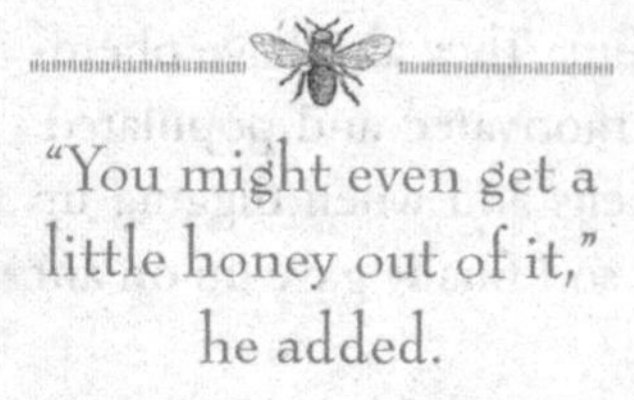

20,000 Bees—Postage Paid

This was getting more complicated than I'd intended. It wasn't like I really wanted to interact with bees. They scared me to death, actually. I'd just wanted to watch them from a vantage point that was safely behind Plexiglas.

In those pre-internet days, finding out what to do next wasn't that easy. I did what I usually did when confronted with a great unknown: I went to the library and found a book on the subject. Bolstered with book learning and absolutely no direct knowledge, I discovered that I could buy equipment and bees from the only retail establishment foolish enough to issue me a credit card, Sears, Roebuck & Co., back when it was a retail powerhouse. Besides its huge general catalog, it issued a dozen specialized ones, including *Farm & Garden.*

I was making $8,400 in 1980 and I'd already spent about $100 on an empty observation hive. A beekeeping starter kit, complete with a build-it-yourself hive body box, smoker, "sting-resistant" canvas gloves, book, and protective veil cost maybe $60, and three pounds of live bees with queen, another $30. This was becoming a very expensive whim, but I was halfway in and I couldn't quit because an empty observation hive would be as depressing as a dead ant farm.

Luckily, the kit came first. It gave me the chance to put everything together, read the book, try on the frighteningly flimsy gloves and the gap-prone veil. With it came a notification that the bees would arrive in a few weeks when the weather warmed up a bit. "No hurry," I thought.

In 1980 a beekeeping starter kit, complete with a build-it-yourself hive body box, smoker, "sting-resistant" canvas gloves, book, and protective veil cost maybe $60.

Weeks passed. One morning, at 5:30 a.m., the phone rang, waking me from a sound sleep. The voice, shrill and loud, cut sharply through my drowsy fog. "This is the Berkeley Main Post Office. You have to come right away. There's a box full of bees over here. They're waking up and they're buzzing really loud."

"Are they loose?"

"Not exactly. But the box isn't secure and they could sting somebody. Get over here right away!"

Going Postal

I was mystified, but I grabbed my gloves and veil on the way out, just in case. Not that I really knew what I'd do if there were a problem. As instructed, I pulled around to the loading dock and explained why I was there. Looking relieved, the sole worker back there went inside and came out wearing gloves and gingerly holding the outer edges of a small wooden box. It was buzzing. The two largest sides were covered with screen and I could see a mass of bees inside hanging in a large mass. I noted that the screens were doubled in a way that the bees' stingers wouldn't reach, even if he were holding the box from the sides. I relaxed a bit.

The book laid out the steps for what to do from here. I had memorized them, not wanting to be paging through it later, covered by bees in my "sting-resistant" gloves. The next step was to put them into a cool, dark place until just before twilight. When I got to school I put them in a closet where the bees could rest, except during every class break when curious kids came in to sneak peeks at the buzzing, slowly writhing, mass.

I was excited. I was scared witless. I tried to imagine the next steps: gently reaching into the center of the mass and extracting the queen cage, a small screened box, removing the cork on one end, hanging it from one of the frames inside the hive, and then shaking and banging the box until all the bees had fallen into the hive. Put on the lid, put some grass in front of the hive opening, and walk away.

At the appointed time, I put on my gloves and veil and carried the box in both hands at arm's length up to the platform where the hive was already waiting. I was followed by a small group of curious friends and faculty members. They huddled some distance away as I went through the checklist of the next steps. ("Don't forget the cork!")

With that kind of sitcom-like setup, you'd expect that everything would go disastrously wrong. Sorry to disappoint: As the sun went down, I followed the directions. Things went like clockwork. I remembered to remove the cork. I opened the box and the bees didn't fly away or sting me silly. I shook and jostled them into the hive and they spread over the frames, seemingly relieved to be home among beeswax again. I closed the lid, placed the dried grass in front of the entrance, and felt competent and brave and alive, like I was somehow home again as well.

Ignorance Can Kill

The next day, things were fine. It was a warm, sunny day, and when I went up to the bee platform at lunchtime with some students, the bees were flying experimental flights out of the hive. Things looked good.

My sense of competence was short-lived, however, and I felt terrible about what happened next. My book-learned beekeeping somehow failed me, or maybe I didn't read far enough into the guidebook.

Not long after, the weather turned cool and wet again. One of my students hurried down from observing the bees to report that they were acting weird, climbing out of the hive and walking listlessly around on the deck. He reported that some seemed to be dead or dying. I climbed up to the platform and looked on helplessly for a few minutes, trying to understand. I began desperately paging through the disease sections of my reference books, wondering whether this was a sign of something like foul brood or deformed wing virus, or whether they'd been poisoned by the bee-unfriendly buckeye trees nearby.

I opened the box and the bees didn't fly away or sting me silly. I closed the lid, placed the dried grass in front of the entrance, and felt competent and brave and alive . . .

It took a while, but I finally realized that they were starving. They were starting a colony from scratch and needed a lot of food to build the comb. The book recommended something I had skipped over: that they should be fed until they get established. They prefer honey, but can live on sugar water. When the cold wet snap hit, they were mostly housebound at a time when there was nothing in the cupboards.

Chastened, feeling like a helpless incompetent who shouldn't be allowed to supervise bees, or even school kids, I had to get back to class, so I put a few of my students to work.

Happy to miss a few minutes of their class for a worthy (or frankly, for *any*) cause, they mixed up some sugar water, one part hot water to one part sugar. We poured it into a feeder (basically an upside-down mason jar with nail holes pounded into its lid), and took it up to them. Even before we attached it to their hive, the bees smelled the sugar and began streaming excitedly to the jar. Most recovered from their lethargic stupor. Within a day, they had emptied the pint jar and were asking, "Please sir, may we have more?" We continued providing the sweet stuff until the weather cleared up for a long stretch and they began turning up their noses at it.

Still No Observation Hive

I'd like to be able to report that my new beehive quickly became a healthy, fully functioning hive that allowed me to finally do what I'd originally intended to do—set up an observation hive—but it turns out that it doesn't work that way. The bees had a lot of work to do before they'd even have a viable hive, much less be able to donate precious comb and bees to a new hive. It would be days, maybe weeks, before they'd build some of the comb deep enough for the queen to begin laying eggs. Those newly laid eggs would take another twenty-one days before they'd emerge as adult bees. During honey season, a bee may live only forty-two days, which means that a large percentage of the mail-order bees would be dead before the first batch of new bees would replace them.

In the best case, it was clear that by the time this hive would be able to donate combs to my observation hive, the school year and most of the honey season would be over. It became clear what I needed to do: buy another hive, one that was already established, healthy, and fully functional. It was time to call that beekeeper back and see if he'd sell me one of his beehives. The quest for an observation hive was becoming more and more expensive.

40,000 Riders—Carpool Lane?

I had never even seen a fully functioning beehive before going to pick up my first one. The bemused beekeeper I'd talked to earlier agreed to sell me one. As instructed, I arrived at his home at sunset. Not that there was anything fly-by-night going on. In fact, the exact opposite was true: field bees *don't* fly by night, so waiting until dark makes sure you don't leave any behind.

Before my wife and I arrived, I was concerned about how I was going to move the hive fifteen miles to its new home at the school. I had already read a few bee books, so I knew that the pieces normally just stay together by the miracle of gravity, but that it's possible to bind them together to transport them. I was relieved to discover that the beekeeper had already done that.

He had blocked the entrance with wood and tied the layers of the hive together with wires and straps.

In retrospect, I don't think I'd really thought the next part through too well. The beekeeper came out and looked dubiously at the vintage VW Bug in which we arrived. As much as I'd thought about the actual transporting, I had not really thought about how the VW was built. I had arrived with the vague expectation that a beehive would fit in the trunk but if it didn't quite fit we could put the hive in and tie the trunk shut. In a normal car, that would've probably worked, but in the excitement of buying a beehive and going to pick it up, I had forgotten two things: how very small the Bug's trunk is, and where it's located. It's in the front of the car because the engine is in the back, which means you can't just tie down the lid and still see where you're going.

My wife and I looked helplessly at the hive, at the car, at each other, and back at the hive. "What have I done?" I thought. "And what do we do now?" It was a nearly singular experience of excited helplessness (one I would experience again a few years later when we got home from the hospital and put our newborn baby on our bed and realized that neither one of us had any idea what to do next).

I checked the straps. They were tight. I listened to the bees. They seemed to be settling in for the night, not particularly excited or upset, but what did I know? I had had no experience with beehives and had only the thinnest veneer of book learnin' about them. But here we were in a VW Beetle, and there was only one place the bee-filled hive was going to fit. I put it in the backseat behind the driver's seat, strapped a seatbelt around it, checked the ties a few more times, and started the car, reviewing everything I could remember about the principles of "defensive driving" from driver's education a decade earlier.

I tried not to think of what would happen if we had an accident. I tried not to think about the fact that at the other end I'd have to carry the forty- or fifty-pound hive in the dark up a hillside path about 150 yards, then wrestle it ten steps up a ladder to a round platform—an abandoned attempt of years earlier at building a raised dome classroom—where the bees were slated to go.

Believe it or not, we made it. I got the hive in the proper place. I removed the straps and, eventually, the block of wood that sealed up the entrance, half expecting them to fly out and attack in the darkness. I even turned my flashlight off and listened for the sounds of angry, vengeful buzzing. When I didn't hear any, I climbed back down the ladder and returned with a handful of grass to leave in front of the opening. The idea, according to my books, is that the grass offers a visual cue to the bees that they're in a different place, so that they'll circle around, noticing the navigational details and reorient themselves to their new home. I still don't know if that's necessary, because I've never moved a hive without doing that. I still do it every time.

Bee Deviled

Bees, you may have learned, are dying from parasites and diseases more than ever. According a March 29, 2013 article in the *New York Times*, for decades it was considered normal for individual beekeepers to lose five to ten percent of their hives. Since the 1990s, though, that percentage began edging upward each year, and now beekeepers typically report that half or more of their hives are dying each year.

What's killing the bees? Everyone has their own pet theories, based partly on their own biases against certain things; and there's good reason to believe it's a combination of things: the constant moving of hives by commercial beekeepers; several diseases and parasites; and (according to many scientists and the European Union, which banned it) neonicotinoid insecticides.

However, everyone acknowledges one big cause, varroa mites, so destructive to bees that in 2000 it was even given a scary name as its scientific nomenclature: Varroa destructor.

V. destructor is a nasty bit of work and a fairly new one, believed to have evolved to its present state of destructiveness in the 1960s in the Philippines. The tiny pest, nearly invisible to the naked eye, spread across the world over the following decades leaving destruction in its wake.

It hitches rides from one colony to another by hopping from bee to bee. It bites into its host and sucks its blood, often jumping to a nurse bee when it gets to the hive, where it feasts and waits until a bee larva is about to be completely sealed up in its final stage of development into an adult. If they jump in too early, the mites can be discovered and removed, so they wait until the last second before making their move.

Sealed in with the helpless larva, multiple varroa mites feast, mate, and lay eggs. They make gashes on the larva's body, which act as a gateway for a number of parasites and diseases that usually prove fatal to the larva and, eventually, possibly to the entire hive.

What to do? Some large scale beekeepers say, "Nuke 'em with pesticides!" That worked for a short time, killing ninety-five percent of the mites. Unfortunately, the five percent that survived quickly became a population that is now immune to that pesticide. There aren't many pesticide alternatives left because most chemicals that kill mites will also kill bees.

There are hundreds of native bee species that do the same job but have been mostly crowded out of settled areas by domesticated honey bees, in the same way cattle has crowded out elk and buffalo in many places.

More natural methods that have been tried with claims of success include powdered sugar (the fine grain and cornstarch dislodge the mites from the adults), formic acid, essential herbal and citrus oils, increasing heat within the hive, increasing cold, and breeding "hygienic" bees that will pick the mites off each other and dispose of them outside. Weirdly, the last one may be the most likely successful method, and so some beekeepers don't treat their hives with anything at all, figuring that the most "hygienic" bees will be the ones that survive and eventually breed mite-resistant bee populations.

I hope that latter theory is true, and I see some evidence of an individual hive here and there being more resistant to damage than the hives on either side of it. I do not intend to resort to pesticides.

THE SMARTEST THING EINSTEIN NEVER SAID

Maybe you've heard this quote attributed to Albert Einstein: "If the bee disappears from the surface of the earth, man would have no more than four years to live. No more bees, no more pollination—no more men." There is no evidence that Einstein ever said this or anything like it. He was a physicist, not an entomologist, after all. The first known appearance of this "quotation" appeared in 1994, long after his death, in a pamphlet from French beekeepers demanding higher tariffs on foreign honeys.

It appears to be one of those quotes manufactured by activists and attached to a famous name to give it more credibility. It's sort of true. Many plants depend on bees to help pollinate them. But they don't have to be *honey* bees. There are hundreds of native bee species that do the same job but have been mostly crowded out of settled areas by domesticated honey bees, in the same way cattle has crowded out elk and buffalo in many places. Those native bees would likely make a comeback if all honey bees died, eventually doing the job of pollination.

Guilt about Indigenous Bees

I do have to admit to some sadness about native bees. When I first moved to the isolated end of my home island nearly a decade ago, I noticed that I could find no honey bees here. Flowers didn't buzz with that familiar honey bee sound, but they *did* buzz with other sounds: huge loud, black bumblebees, for example, and quiet, small bees with colorful, clown-like, stripes.

Nowadays, with my hives and those of a community garden nearby, I notice honey bees near most of the flowers I pass. The other bees? Not so much.

UNSTOPPING A JAR ON A JANUARY EVENING

"Whose jar is this, I think I know . . ."

It's mine, actually. That ownership was somewhat up in the air for a while, because I had saved two jars from the autumn harvest, knowing from experience that before the spring harvests somebody would likely ask me for honey with some urgency. In retrospect, it seems strange that I didn't save any for myself, but I consume so much during harvesting that I can lose the taste for it for a while. So, for nearly three months, I've had two jars in my cupboard, packaged up pretty, waiting for the day.

Sure enough, my daughter asked me for a jar just before Christmas for her very deserving mother-in-law. No question about that being a good thing, and so it came to be that Jar #1 found its home.

New Year's came and went, and I began thinking about the jar in the cupboard. It occurred to me that maybe I had found a home for Jar #2. Why not me? Not that I needed honey, exactly. I just started thinking about why I often don't save any honey for myself. I was missing the smell and taste of it, both on its own terms, but also as a way of reliving the joy of warm weather and sticky harvesting, of warmth and the feeling of abundance that comes as pint after pint of honey emerges to fill another mason jar.

I removed the pretty packaging, unscrewed the top, put a spoon in and again tasted the autumn, the sunshine, the flowers, and the sugar rush. I replaced the lid and placed the jar back in exactly the same cupboard, but this time feeling one jar richer than before. As a bonus, I also felt relieved of the burden of deciding who should get the remaining jar. No more "promises to keep, nor miles to go before I sleep."

Less than Infallible

I've gotten a lot of expert advice over the years. Lots of it has been valuable. Lots of it has been reasonable but completely contradictory to the other advice. It's a wry old bit of humor that you get three beekeepers in a room and you'll get four different points of view on almost any issue. I learned that early. I'm a researcher by trade and temperament, so I don't mind cracking books, asking questions, reading websites. In most cases, you start seeing where the advice overlaps and where it doesn't. There's some actual research, but most of the knowledge of any beekeeper, myself included, is a mix of received wisdom, observation, anecdotal conclusions, half-baked theories, and unproven certainties.

At first, as a beginner, I was looking for the "one right way" to handle bees. By the second or third book, though, the conflicts started causing me a lot of anxiety as I tried to make them into a coherent whole that I could follow. From there came the disillusionment of trying to determine whether any of the authors really knew what they were talking about. I began despairing slightly about whether I'd ever be able to master this.

At some point, though, the conflicts among experts freed me. Getting stuck eventually led to a conclusion: "Maybe there isn't just one way that works."

I began tallying the things the authors all agreed about. For the short term, at least, I'd accept these things as probably true. Where they disagreed was where I was free to think, experiment, and figure out what seemed to make the most sense, but even some of that I discarded. For example, I had hoped that bees could be raised naturally, organically, without using a bunch of chemicals to keep them healthy. The books weren't promising. Each had a chapter about serious bee diseases and the chemicals that should be used to treat them. Eventually, through experience and seeking out new information, I found that it was possible to raise bees without that stuff.

I learned that there was an awful lot that could be learned from books, but that it wasn't enough. I, so analytically focused on problem-solving, had to stop, listen, watch, and learn from the bees themselves. I can't train them to do their jobs, but they've done a fairly good job of training me to do mine.

Stupid Buckeyes!

Having grown up in Michigan, I knew all about the football rivalry with Ohio colleges, so I got used to hearing Ohio Buckeyes being regularly disparaged. I was surprised, though, to be among Californian beekeepers who likewise badmouthed those "damned buckeyes."

It turned out, though, that they weren't talking about football teams, or even Ohioans—they were talking trash about buckeye *trees*. It turns out that some buckeye trees can poison bees and destroy colonies. Buckeye pollen can cause developmental defects, resulting in larvae developing into adults without wings. If enough bees in a colony develop that way, the damaged bees can't forage, and the colony starves to death.

The good news is that bees prefer the pollen from other plants if they can find it, and only collect from buckeyes if it's all there is. The bad news is that a number of other plants are also poisonous to bees, including azalea, jimsonweed, plume poppy, and the tea plant.

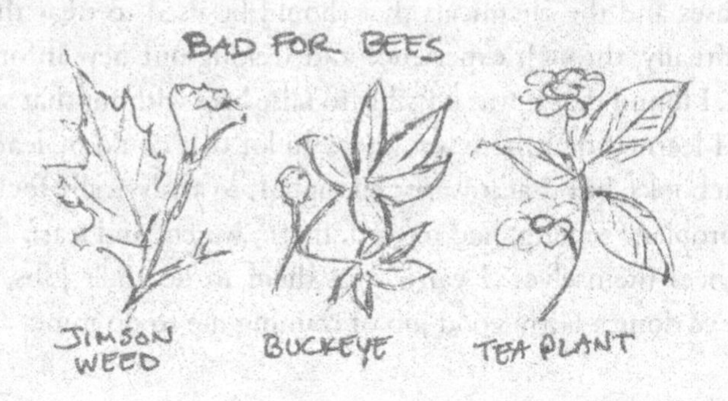

Honey Is Not Bee Vomit

Kids of all ages like to gross each other out. One story they like to tell is that honey is "bee vomit." This is based on a misunderstanding of how bees carry nectar from flowers to the hive. Bees have a special "honey crop" where they store the nectar. It is separated from their actual stomach with a one-way valve. The valve is a pretty cool trick. It allows the bee to transfer some of the stored nectar to their stomach if they start running out of energy on the flight back home, but it prevents the contents of the stomach from reversing direction into the crop. So it's not bee vomit. Tell your kids.

DEFYING GRAVITY

A foraging bee can carry back to the hive about 40 mg of nectar, which is pretty impressive considering that she herself weighs only about 120 mg. Equally impressive is that worker bees inside the hive can carry about 100 mg of nectar on foot.

Can't Fly a Straight Bee Line

Bees sometimes collect juice from fermented fruits, and, yes, they can get tipsy. Also, honey that hasn't dried quickly enough can ferment, creating mead that can likewise knock bees off their feet. It turns out that the effect of alcohol on bees is very similar to its effect on humans. They wobble when they walk. They have more flying accidents. They become more aggressive. They have more trouble learning new things and finding their way back to the hive. There are a few things that are definitely different (intoxication from a single binge can last as long as forty-eight hours) and arguably different (drunk bees stick out their tongues more often than sober bees).

It turns out that the effect of alcohol on bees is very similar to its effect on humans. Drunk bees stick out their tongues more often than sober bees.

Balancing the As, Bees, & Cs

Beekeeping balances out much of modern life. Too much of our daily lives are ambiguous: the fluidity of friendships, the ever-changing status of being a partner, a spouse, a parent, a wrangler of teenagers, or a worker within a job or organization in which the goals of the daily job are often unclear and amorphous. Bees in a hive are largely binary. They are either there or not, alive or not, producing honey or not. There's little that's postmodern or ironic about building hive bodies, watching the bees do their work, harvesting honey, or filling jars with honey. I love that.

How Great Thou Art, Reverend Langstroth!

Reinventing the beehive has been an impulse of beekeepers for centuries, and bees have managed to tolerate most of their tomfoolery, more or less. The beehive that I use has been the standard for most beekeepers worldwide for more than 160 years. It's the Langstroth hive, patented way back in 1852.

The hive is boxy, but also elegant in a completely utilitarian sense. The elegance of the hive came from an amateur's scientific observation and the desire to do as little damage to the bees and their work as possible. Its inventor was a Congregationalist minister in Philadelphia, the Rev. Lorenzo Lorraine Langstroth, who took up beekeeping as an antidote to severe bouts of depression. Building on the work of others, he discovered the concept of "bee space," which is remarkable in its simplicity. If presented with surplus space narrower than one-quarter inch, the bees will fill it in with propolis, the sticky tree sap they collect for filling nooks, cracks, and crannies; if wider than three-eighths inch, they'll fill it with honeycomb. Of course, honeycomb is preferred. Knowing that, Langstroth built hives like top-hanging filing cabinets, in which frames hung with precise gaps between them, which encouraged the bees to fill combs in the right places and not gum up everything with propolis. That allowed each comb to be lifted out of each box "like pages in a book," meaning less damage and fewer bees killed during hive inspections and honey harvesting.

That said, there are some problems with the Langstroth hive, especially among farmers of the developing world and hipster urban farmers of the post-industrial West. Foremost is that they are relatively expensive, requiring precision-cut lumber shapes and patterned sheets of beeswax. So some people took what looked like several steps backward and began championing the top bar hive.

Top Bar Hives

The main benefit of top bar hives is that they're easy and cheap to make out of whatever materials are available. They resemble Langstroth hives in that the combs hang from supports on the tops of the boxes with "bee space" in mind. They are different in that, unlike Langstroth hives, top bar hives are only one box and are not built to one standard size. The guides for comb building in a top bar hive are not frames with thin sheets of wax stretched across them, as is the case with the Langstroth hive, but just a bar with a small strip or smear of wax so that the bees will (hopefully) build unsupported hanging combs straight across them.

But no matter. The benefits of top bars over standard Langstroth hives are their cheapness, use of recycled material, lightness (only individual frames are lifted, not the wooden boxes that hold them) and their funky-pretty style. Disadvantages, though, are pretty substantial. For one, you don't get nearly as much honey. You also can't add additional hive boxes on top to accommodate a growing hive (often inspiring half the bees to leave the hive in a swarm). Finally, you can't empty the unsupported combs in a centrifuge extractor, so harvesting honey means breaking the comb into bits, which means the bees have to build new combs after every harvest. (Langstroth combs, supported by wood on all four sides instead of just one, can be emptied of honey and replaced unharmed for immediate refilling by the bees.) You might consider top bars as a cheap way to try out the hobby before investing in Langstroths.

HISTORY OF THE TOP BAR HIVE

Top bar hives are sometimes called Tanzanian or Kenyan top bar hives. From this, many Westerners (especially those who treasure the wisdom of faraway tribal societies over their own) assume that they are an ancient invention by indigenous people in those two countries. It turns out that this is not true.

The real story is that the top bar hive was invented in the 1970s by Westerners from the Peace Corps and Canada. Their aim was to end the traditional method of harvesting honey, which was finding a bee tree and chopping it down, destroying the hive and killing the tree. The idea was to reduce the destruction in wildlife preserves by convincing the locals to *raise* bees instead of *hunting* them.

Harvesting Honey

There is nothing sweeter, both figuratively and literally, than harvesting honey, but it is sticky and sweaty work. Some beekeepers harvest only once or twice a year, in the fall at the end of the honey season and sometimes in the middle of the season as well.

I have a different philosophy. I begin harvesting as soon as there's honey to be had, and I do so every few weeks afterward. I harvest this way because I believe it causes less stress to the bees to lose a couple of frames of honey at a time when there is still plenty of nectar out there, with the full combs immediately replaced with empty combs that are ready to fill. I believe that the empty space decreases their urge to swarm and inspires them to collect more honey. Furthermore, I believe that autumn is the most traumatic time for a big harvest, since they want to have full stores as winter comes on.

The evidence for this is that autumn is the time that they are most aggressive in defending their hive and in robbing weaker colonies if they themselves have been robbed of their winter food supply.

There's another reason for harvesting throughout the season. It makes sense commercially. The people who love my honey want to get some as soon as possible after a long winter drought. They also like tasting the differences in the seasonal output, from the light colors and flavors of spring honey through the dark molasses flavors of the autumn. For those who believe that the trace amounts of pollen in the honey help relieve their hay fever, it also makes sense to be eating honey with the most recent pollens of the ever-changing seasons.

Spinning the Truth

Extracting honey with a hand-cranked centrifuge is a little like making cotton candy. Microscopic strands of honey get airborne, covering me and everything within several feet with a subtle layer of stickiness. In a single harvest, I might extract from twenty to sixty frames of honey, two at a time. It's fairly strenuous work and best done on a warm to hot day, so by the end of the day, I am very grateful to shower the sweat and stickiness off my arms, face, and body. I do, however, like expending the physical exertion instead of using an easier, more common, motorized extractor.

The process starts with a very sharp, double-edged, serrated blade. With it I saw off the *cappings,* the thin layer of wax that bees use to seal up honey when it's done drying and ripening in the comb, without doing much damage to the rest of the comb. The combs slide into a wire basket in a stainless steel barrel that's about the size and shape of a small, tapered trash bin; the handle works exactly like the handle on a salad spinner, gears translating each turn of the handle into several turns of the basket inside.

The honey is heavy and the wax is fragile. You have to start spinning slowly at first, extracting only some of the honey from each side of the comb so the weight of the honey on the other side doesn't collapse the delicate structure into pieces. Each time you turn the comb around, you can go a little faster until finally the basket is going maybe three or four rotations a second and the honey is flying in long threads to the sides of the barrel and slowly gliding down. When enough collects, you open a faucet at the bottom into a very fine strainer to screen out chunks of wax, propolis, and other foreign matter, but not so fine that it removes the bits of pollen that slightly cloud the honey, adding a bit of protein (and, perhaps, tolerance to the allergens). With the honey bucket, you beginning filling pint-sized mason jars, weighing them on a scale as you fill them. When you get just a smidgeon more than 22 ounces of honey, you seal it up, label it, and offer it to a honey-hungry world.

HEAVY HONEY

"A pint's a pound the world around," goes an old memory aid for cooks and food servers. However, that's only true for water. Honey is so thick that a pint of it weighs 1 pound, 6 ounces.

I'll Follow the Sun

It almost looks like an act of worship. A bee comes out of the hive, pauses for a moment in the bright morning light, and does the insect version of genuflecting in the face of its god. It doesn't cross itself; it circles itself. Circles itself, that is, around the front and back of the hive in a slanting halo perhaps ten feet in diameter. It is orienting itself, getting right with its God, its Great Mystery, its Guide, its Light and Salvation, the Sun.

Bees use the sun for a lot of things. Just like the rest of the world, bees use the sun for timekeeping in the morning and evening and for the light that reflects in through openings and cracks, making it possible to see inside the hive's darkness. And of course, they use it for heat. When the sun warms the hive enough on a cold day, bees unwrap themselves from the protective blanket of tens of thousands of their peers and get to work, the bee equivalent of waiting for the furnace to kick in before emerging from the covers.

There's another use, though, that's less universal across the animal world: navigation. Bees use the sun as their GPS in the sky, or at least their compass, religiously tracking its position to know what direction they're heading. It's one of the few things outside their hive that they have a symbolic word for in their unspoken language. When a bee does its figure eight dance inside the hive to communicate how to get to choice nectar, every other bee understands that "up" means "sun." That is, if the target is fifteen degrees to the right of the sun, the dance will be angled fifteen degrees to the right of straight up. That's a fairly sophisticated form of symbolic language. Especially for an organism with a brain smaller than a sesame seed.

The sun's position is so important to finding the one true way, out and back, that they even have learned over the eons to anticipate its fluid nature, that it starts a pathway each day from one direction and ends it in nearly the opposite direction, and that its position changes minute by minute. Their flight paths adapt to those changes, with the bees correcting their directional orientation slightly with each passing minute.

When a bee does its figure eight dance inside the hive to communicate how to get to choice nectar, every other bee understands that "up" means "sun."

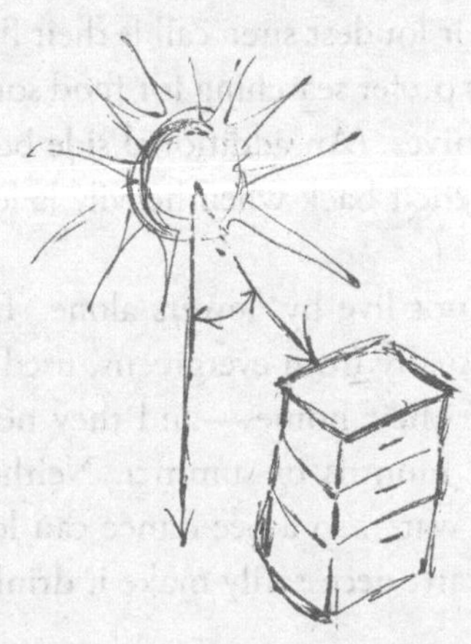

Smell You Later

The sun isn't the only part of bee navigation. A bee can use it to get close to where it wants to go, but once nearby, they have to sniff out the rest of the pathway. That requires an exceptional nose, and luckily the bees have exactly that.

The benefits of flying instead of walking like ants to their food, of course, include speed, range, and safety. However, bees lose one benefit of walking that ants use a lot, a chemical marking smell that keeps them from wandering off the path. There's no way to scent the air and have it stay in place, and marking the ground below their flight path would be impossible.

Luckily, the flowers that want to attract bees have ways of advertising their location, of communicating "hey!" for a bee to see. Of course, shapes and colors of flowers attract a bee's attention, but their loudest siren call is their floral scent, which is one reason bees prefer searching for food sources that are upwind from their hives. (An additional side benefit is that they have the wind at their back when heavily laden with pollen or nectar.)

But bees do not live by flowers alone. They need propolis—sticky sap, usually from evergreens, used to fill unwanted gaps and holes in their homes—and they need water, particularly in the hot months of summer. Neither is particularly smelly, especially water, so a bee dance can lead a bee toward water . . . but it can't necessarily make it drink. Unless a water source is very visible, it's possible for a bee to get within a few yards and not see that dripping faucet behind the geraniums.

Luckily, though, there are stinky bee feet.

Bees can exude a smelly liquid. There's some dispute about whether it's actually exuded from their feet, or from somewhere else from whence it's passed along to the feet, but either way, there's a scent that's left when bees start landing and walking near a water source. It just takes the stinky feet of a few bees to create enough of a foot note leading other bees to the right spot.

Amazing, actually. If only our lives were so simple. If only we could reach our goals simply by sniffing the air for dirty socks.

UNLEASH THE BEES OF WAR

Bee brains, although tiny, have about a million brain cells (also known as neurons). Rats have about 100 million. Humans have about 100 billion. Still, in what it does best, the bee can often beat the mammals with the big brains.

Of course, anything that amazing is going to be weaponized by the primitive monkeys we call humans. British scientists in 2012 began working on ways to simulate a bee's brain in order to then replicate its abilities. They hoped to team up with

continued

American scientists working on creating miniature bee replicas that can spy, unnoticed, on others. Ultimately, though, they hope for more than that. A bee robot that could learn and navigate could also be useful for lots of tasks beyond aggression, for example, flying and crawling into impossibly tight passages to locate survivors of earthquakes, mine cave-ins, tornados, bombings, and other disasters.

Artificial bees would presumably be too small to carry a payload, but perhaps they could be designed to "sting" with something more lethal than bee venom.

In case of a horrifying natural apocalypse or a gross overuse of pesticides, mechanical bees could (presumably at a huge cost per bee) be programmed to do bee things like collect nectar for honey and pollinate our glowing radioactive crops.

If they can pull that off, it will be the only drone that actually looks and flies like the *original* drone (a male bee).

Ironically, and sadly, peace-loving bees have been used for warfare since long ago. A general named Heptakometes in Asia Minor defeated a Roman army in the first century BC by leaving jars of honey from poisonous azalea and rhododendron plants in a narrow mountain pass. When the men of Pompeii the Great stumbled onto the honey, they devoured it greedily, and quickly succumbed to vomiting and diarrhea just before Heptakometes and his men attacked.

Queen Olga of Kiev in 946 used generous servings of mead—wine made from honey—to intoxicate and immobilize 5,000 enemies before slaughtering them. Russians played a similar trick against 10,000 Tatars in 1489.

Long before B-52s and buzz bombs, there were buzzing bee bombs. A succession of armies starting with the Romans catapulted scores of beehives into the camps and castles of enemy soldiers. The enraged bees were said to have turned battles and broken sieges. Fifty pirates in the Mediterranean on a small ship reportedly defeated a much larger ship with ten times as many men by tossing beehives from the masts of their ship onto the bigger ship's deck. Similar bee warfare, sometimes using trip wires to knock over stacked hives, were reported during the American Civil War and World War I.

It just doesn't seem right to use innocent insects in war. All we are saying is give bees a chance.

I WISH I HAD A HONEYGUIDE BIRD

Around 2.5 million years ago, one theory goes, humans in Africa discovered that honey was a concentrated form of food energy unlike anything else in their natural world, providing enough calories and proteins to spur brain expansion over many generations. Long before beekeeping, there was honey hunting, chopping down bee trees and stealing the honey and larvae for food.

Some indigenous tribes in Africa still do this, as do our closest relatives, the chimps and bonobos. (This, incidentally, might explain why African bees developed the habit of ultra-aggression when threatened, as opposed to domesticated bees in Europe and Asia that were bred for docility.)

That's where the honeyguide birds come in. Over the years, they've established a rather amazing symbiotic relationship with humans and other honey thieves.

Honeyguides, found in Africa and Asia, love eating the wax and larvae from beehives. They are also very good at locating beehives. Only one problem, though: they're unable to open up hives themselves. So, they perch near beehives and wait. Whenever they see a person or animal that might be interested in raiding for honey, they whistle furiously to attract attention to the spot. People and honey badgers are their most reliable customers—willing and able to bust open a beehive, but wasteful enough to leave a lot of spoils for the honeyguides. If things go well, everybody wins . . . except (of course) the poor bees.

Bad Honey

Why would anybody screw up the natural goodness of honey? "Money, honey!" of course. It's really a travesty. Testing of honey from major United States supermarkets in late 2011 discovered that most of it wasn't legally honey. *Food Safety News* paid for tests of 60 brands and discovered that 3/4 of the jars contained, at best, honey that had been illegally "ultra-filtered" to remove every speck of pollen.

That doesn't sound so bad, but it is. For one thing, honey naturally comes with trace elements of pollen that add flavor, nutrition, and maybe protection from allergens. More importantly, though, is that "ultra-filtration" was invented solely to commit fraud and sell unsafe honey.

Testing of honey from major United States supermarkets in late 2011 discovered that most of it wasn't legally honey.

How so? Analyzing the honey's trace amounts of pollen is the only foolproof way for regulators to determine a honey's place of origin. Regions of China and India have consistently produced honey containing heavy metals, pollutants, antibiotics, and other chemicals not approved for human consumption, so their honey has been embargoed. By heating the honey to a high temperature and forcing it through ultra-fine mesh, the honey can be shipped around the world and relabeled as coming from some other country without leaving any incriminating regional pollen traces.

Selling ultra-filtered honey is illegal, but the law has not been well enforced, and the banned honey is sold at bargain basement prices, so a lot of honey packers would rather pretend they don't know what's going on.

But at least it's honey, even if it's been overheated and polluted, right? Well, not necessarily. Some of the "honey"—mostly from China, so far—has never seen the inside of a beehive, but instead was created from much cheaper malt compounds and sugar water.

Yeah, it's a little self-serving for me to suggest, but if you want real honey, get your own hive, or get to know a local beekeeper and expect to pay more than you've gotten used to when buying plastic bears full of suspiciously crystal-clear honey in supermarkets.

BY THE WAY . . .

A scientist who analyzes the purity of honey by checking its pollen is called a melissopalynologist.

Honey from Mars! The Mars Candy Co., That Is

I'm often ashamed of my fellow humans for the way they corrupt good and natural things into something appealing, but not good and not natural. Sometimes I'm also a little ashamed of bees for being so easily corruptible by these products.

Let's talk about sugar water. Bees like it. Or at least they like it better than starving. Beekeepers starting a new colony will often offer sugar water to the bees, since the bees haven't yet stored honey and it takes a lot of food to make wax and mold it into combs. If early in the season, it's a good idea, and the bees will take the sugar water for a few days or weeks until they can get a store of more wholesome nectar they can eat. Nectar is the adult bees' natural food, the one they should always prefer, and usually they will stop eating the sugar water as soon as they can, so there's little worry that it'll get into the honey that people buy.

That's why it's alarming to sometimes read news stories of beekeepers finding artificially red or blue or green honey in their hives. In the case of the unnaturally red honey, it was beekeepers in Brooklyn who suddenly started finding it in their hives. It turns out that a nearby maraschino cherry factory kept its windows open on hot days, and the bees got into the syrup the factory used for coloring and sweetening natural cherries into those things you put on top of sundaes. A *New York Times* article posed the disturbing question, "If the bees cannot resist . . . what hope do the rest of us have?"

Even more disturbing was the case of the red, green, and blue honey, in that it took place in France, where they are rather particular about their food and have long resisted the Americanization of their cuisine. In 2012, dozens of beekeepers in the area around the town of Ribeauvillé (Alsace) traced the horrifying honey colors to waste from a Mars Candy Co. plant that made M&Ms. Honey from Mars? *Sacre* red, green and *bleu*!

12-Skep Program

In the 1950s and early 60s, a "beehive" was a popular hairstyle in which the hair was piled and teased to create a dome on top of the head, similar to a Marge Simpson do (but not as tall, of course). The point was to mimic the shape of what we consider the classic beehive. The still popular "beehive" shape comes from something that is all but extinct in modern times, the skep. We know from ancient art that skeps have been around for at least 2,000 years, sometimes woven from wicker or coils of straw and held together with dried reeds. To make them weatherproof, skeps were often covered with a coat of fresh dung that hardened into a sturdy coating like fired clay.

Skeps did the trick for a long time, and a small number of beekeepers around the world still use them, but they've fallen out of favor and are actually illegal in most American states. The reason is that unlike more modern beehives, you can't inspect the bulk of the combs for disease or mites. That's because the bees actually build their combs on the tops and sides of the skep, making a solid mass, instead of modular, removable frames.

Not only can you *not* inspect the hives, but it's also hard to harvest honey without doing often lethal damage to the colony. In fact, for many centuries most beekeepers didn't even try to keep bees alive after the honey season. They'd kill them all with burning sulfur fumes, extract the honey, wax, and dead bees, and start over again with a new collection of bees the following year.

It's hard to imagine doing that. I guess if you're part of a farming culture, in which slaughtering animals is a regular and expected occurrence, slaughtering bees also makes sense, especially if you're housing them in something where you can't easily harvest the honey without doing harm (and probably get stung many times in the process). Still, after years of trying to keep bees alive, of being careful to leave them enough honey to get comfortably through the winter, the unnecessary waste and carnage is incomprehensible to me. But then again, I admit to having a tender heart and squeamish stomach.

. . . for many centuries most beekeepers didn't even try to keep bees alive after the honey season.

HOW TENDER IS MY HEART?

I once drove three miles back across town when I'd discovered that one of my youngest bees, not able to fly yet, had hitched a ride on my pants leg and wouldn't be able to find its way home.

Call Me Five Eyes

"Ocelli" is a strange word. It sounds like an instrument in an orchestra, but it's less important to think about what ocelli sound like than what ocelli *see.* An ocellus (from the Latin meaning "little eye") is a primitive light sensor, and bees have three of these on the tops of their heads. This is, of course, in addition to their two big compound eyes that cover a lot of their face, like sunglasses on a fat Southern sheriff in a 1970s movie. Compound eyes are great for bees in that their many facets provide the widest possible range of vision in all directions. You'd think they'd be enough, but they aren't.

On the face of it, the ocelli don't make a lot of sense. They don't focus and are only good for registering gradations of light, like a photoelectric cell or a very foggy window. Simply registering light turns out to be a good thing though for keeping track of where the sun is. This is important for two reasons. First, it keeps the bee attuned to which direction is up, which isn't always clear in the dizzying world of flying. Secondly, the bee can navigate with the ocelli alone, saving their big compound eyes for tracking danger, obstacles, landmarks, and flowers.

ODD NUMBER?

Why does a bee need three ocelli? Well, the top of its head is pretty severely rounded. One ocellus is forward-centered to catch light coming from directly above and ahead. The other two, located on each side of the head, catch light beside and behind the bee.

Sympathy for the Hexopods

Can you imagine what it must be like to have six legs? I can't. I literally cannot.

I have a pretty good imagination, and a pretty good sense of empathy. It's easy for me to imagine being a four-legged animal, for example, or a bird (because who can't imagine flapping our arms and flying?) I can even imagine how it would be to have a tail like a cat's to counterbalance movements.

But being a bee? They have one pair of feelers (antennae), one pair of multi-functional front legs, two pairs of walking-only legs . . . and also two pairs of wings. Unlike our four appendages, they have twelve. Somehow, even with tiny brains, they can make them all work yet still have enough brainpower left over for complex social activity and navigation. Amazing.

BEE FOOT FETISHIZING

As a kid, I assumed insects like bees and houseflies managed to land on walls and ceilings with some sort of glue or suction cups, and dreamed of having something similar. Luckily, my attempts at inventing something similar didn't do (much) damage to me or the walls of the family home. It turns out that I was going about it mostly all wrong.

It's true that bees, like houseflies and many other insects, *can* exude a sticky substance from their hairy footpads to help them walk on especially slick surfaces. But that's on exceptionally smooth surfaces, like glass; most of the time they don't need to do that. What they do on hard surfaces is exactly what a human wall walker would do: find cracks, crevices, pits, and hollows that will make a decent toehold. Except, of course, bees don't exactly have toes; at the ends of their legs they have pointed claws that can find a foothold on microscopic irregularities. On flexible things like leaves and people, these tarsal claws can hook and hold like miniature cat claws.

If you look at almost any surface with a well-lit microscope, you'll likely see that most have a wealth of toeholds that would hold you . . . if your toes were pointed and you were the size and weight of a bee.

Blind Spot

We all have our blind spots. Bees are no different.

We can never actually see our own backs, only reflections in a mirror or a photo. Even weirder, there's a part of our backs that most of us have never even touched, making it the perfect place for some joker to tape a "kick me" sign without us knowing.

Bees have a similar spot, and nature often sticks a sort of "kick me" sign right on it.

Bees are incredibly clean insects, as befits their job as food handlers. It is not just their aloofness that gives them the reputation of being the cats of the insect world, but their continuous self-grooming. They use their front and middle legs as brushes to remove the pollen dust that's a constant in their work environment. Their front legs even come with a specialized notch that in placement, size, and shape appears to have only one function, described by its name: the "antenna cleaner."

Given all that, it's almost comical to sometimes see nature's kick-me sign on the back of incoming bees. Just above their middle pair of legs, in a spot they can neither see nor reach, they sport a bright white, yellow, orange, or red streak. It's pollen from the flowers they've visited, making the bees look like a family of tiny flying technicolor skunks. They'd be so embarrassed if they knew.

DUST MAGNETS

A big reason that bees make great pollinators is that their fuzz has an electrostatic charge. They get that static electricity from rubbing up against other bees in the same way you get it from petting cats or rubbing a balloon on your head. As a result, they get covered from head to tail with pollen that they unwittingly spread from flower to flower. What happens to the pollen they brush off themselves after visiting multiple flowers? It doesn't go to waste—they stuff it into "pollen baskets" on their legs and carry it back to the hive for the larvae to eat.

BFF
(Bee Friends Forever)

Don't get me wrong, people are great most of the time. But sometimes they are just too damned unpredictable, secretive, spiteful, moody, vengeful, incomprehensible . . . add your own adjectives here. Those are the days I hang out with my bee buddies. We've got a straightforward relationship. I shelter them and they nourish me. (Literally, as in honey for my breakfast cereal, but in less concrete ways, too.) They aren't always predictable, but they're dependable. I have no worries about their feelings toward me—because they really have none. They defend their homes, and if they try to sting me, it isn't personal or vindictive, it isn't because of their personal baggage or projections, it isn't subtle or twisted or dependent or incomprehensible. I've learned to love and respect that about them.

I love that they really don't require anything from me. They're self-sufficient and independent. I want to be more like them, and I want to find more friends like them: symbiotic but not codependent.

Bees Here Now

Bees live in the present. If there is danger, they defend. If larvae need care, they provide. If food or water are needed, they get it. If cold, they gather in a large group; when it gets too hot inside the cluster, they rotate out to the outside; when they get too cold, they burrow back toward the center. They see what needs doing and they do it.

That role model isn't always right for all of life, of course, but it's more right than I'd ever have imagined. So much of my life has been taking note of things that need doing in the future. Doing them *now*, as I notice them, saves a lot of energy and time. Chop wood, carry water, write the book, fix the door, pay bills—don't wait, don't worry, do it now. Sort of a bee form of Zzzzen.

Chop wood, carry water, write the book, fix the door, pay bills—don't wait, don't worry, do it now. Sort of a bee form of Zzzzen.

The Rise & Fall of an Empire

I sometimes think of the great empires of the world while watching the beehives. The thing about successful empires is how strong and permanent they seem to be. The Aztec, the Roman, the English, the American—each in its day was such a juggernaut that it appeared eternal. Yet, each in turn had its beginning, its ascension, its successful run, and its inevitable decline.

And so it is with a successful beehive. It is formidable. It mounts a strong defense against any aggressor, keeping its borders safe from other bees and even animals many times its size. On a summer day, it is a literal beehive of activity, sending envoys out in all directions and bringing back wealth, outperforming other colonies by sheer strength and numbers.

Yet, as with empires, history tells us that the mighty will ultimately be laid low. In both cases, empires or beehives, you may see dozens of small warning signs. Sometimes there's something that can be done to forestall it. Sometimes there isn't. It could be that, as with all living things, there is a lifespan that can be optimized, but not extended indefinitely. But ultimately all things dwindle and die, even thriving empires and beehives. And so it goes.

THE MACK TRUCK OF MORTALITY

Sometimes I worry about what's to become of me. I wish I could say that the bees comfort me. Well, they do. But not enough. I'm scared of dying like a bee, perishing alone when I'm no longer useful. Bees don't have Social Security. Bees don't have health plans. It seems sometimes as if there's a malignant strain of philosophy that wants human society to become more like a beehive. Unless you're part of the aristocracy, you're expected to work yourself to incapacity, then crawl away to die alone.

Live productively, keep flying, and leave an impressive splat on some trucker's front window.

I'm afraid of living on into a feeble decline. My goal is to keep flying as long as I can. When I start slowing down, I hope to be smart enough to look as tasty as I can to a hungry blue jay, or—in a pinch—the windshield of a fast-moving truck. For that's what I've learned from the bees about death. Live productively, keep flying, and leave an impressive splat on some trucker's front window.

Even Bugs Get Bugged by Other Bugs

Day and night, hives are a pretty hectic place. While bees are social insects, they live in an enclosed space without traffic lights or dividing lines, meaning every second is filled with bees brushing by and maneuvering around each other. In fact, it's the primary way that the scents of the queen affect the hive. Her pheromones are continually spread as bees slide, collide, and rub against each other in a mass of pedestrians and traffic jams that make Calcutta look sparse in comparison.

I can't imagine that kind of stimulation and interaction without needing a way to get apart for a while, get alone, get some peace, quiet, and sleep. Of course, bees are not people, but they need the same things now and again. But how? Where do they go when they want to be alone?

Well, if you ever look carefully inside a hive, you get that question answered pretty quickly. You'll see the rear ends of dozens of bees barely sticking out of individual cells in the honeycomb. Tired, stressed, and overwhelmed bees find an empty one and crawl headfirst into it for a little peace, privacy, rest, self-reflection, meditation—whatever it is that bees do. It's a snug fit, but I imagine being inside a cell is a heavenly respite from the bustle and noise that normally surrounds them.

I completely empathize. I sometimes wonder if there's something wrong with me that I regularly need separating from other people at times. But then I look at the sleeping bees, comforted by this realization: They're maybe the most social of the social insects, programmed to interact constantly with thousands of others . . . and yet even *they* need to be alone sometimes.

LIVING IN THE BEEHIVE STATE

Mormons have had a longtime affinity with the beehive. Historians say that Joseph Smith borrowed the beehive symbol from the Masons, of which he was a member. Wherever he got them, there are beehives depicted all over the Salt Lake City temple and headquarters. In fact, when the Utah territory applied for statehood, the Mormons in charge wanted to call it Deseret. According to Smith's *Book of Mormon* "deseret" was the word for bee in the language of the Jaredites, a tribe of Babylonian Hebrews led by brothers Jared and Mahonri Moriancumer after the fall of the Tower of Babel. The Jaredites allegedly traveled to ancient America by barge and lived on the shores of Lake Ontario. Mormon tradition has it that they had hives full of "deserets" thousands of years before honey bees were (re?)introduced to America by European settlers in the 1700s.

The Mormons, like the Masons (and perhaps the Jaredites), believed that the honey bee society was a great role model for their communities. An 1881 editorial in the *Deseret News* explained, "The hive and the honey bee form our communal coat of arms . . . It is a significant representation of the industry, harmony, order and frugality of the people, and the sweet results of their toil, union, and intelligent cooperation."

In a sense they were right. Honey bees work for the good of all, don't accumulate personal stores, but share all as needed, take care of their young, and all that. However, if taken too seriously, you really wouldn't want to live in a society based on the honey beehive. (See the following "Bees Are a Bad Role Model for Communities" section.)

Despite the petition from the governor of the territory, who also happened to be Brigham Young, spiritual leader of the church, Deseret was rejected as a state name. Federal authorities believed that using that name would be stepping all over the "establishment of religion" line. The territory was admitted in 1896 as Utah, honoring the Ute people.

The Mormons managed to establish the state's nickname as The Beehive State. This, despite bees finding much of the state inhospitable. Honey is not exactly a huge crop there. But hey, there were never any wolverines living in The Wolverine State either.

Bees Are a Bad Role Model for Communities

Women do all the work. Only a few men are tolerated long enough to service the queen, but they often come to a bad end.

The workers are worked to death. When they get old, sick, or injured and no longer can do the job, they don't get to live out their days in cushy retirement, enjoying the fruits of their long labors—they get pushed out the front door, without even the opportunity to say goodbye to friends or clean out their desks. If they don't take the hint, one of the workers will pick them up and, barely clearing the ground, fly them a distance and unceremoniously drop them to the ground. If they're lucky, they'll get picked off by a bird and die quickly; otherwise, they'll succumb by the next morning, alone and not comprehending how, after all that work, their lives came to this.

The leader has sex with multiple partners; the workers aren't allowed to.

The leader can live up to ten years, but there's a good chance that she'll be overthrown from her position as hive leader long before then. If the workers, her own daughters, come to believe that she's over-the-hill, she'll be stung to death and replaced.

Bees Don't Talk Back

I talk to my bees. Not all the time, but I do when I work them, when I open up a hive and interact directly and invasively. You'll be (perhaps) happy to know that they don't talk back.

Why do I talk to them? Not because I think they understand. Not even in the way that some people talk to their plants, with the hope that the plants somehow sense what they're saying and will respond to them in some form of mystical life-force communication. True, leaning in close to your corn and whispering into its ears will provide carbon dioxide and water vapor, which plants like. But that doesn't work for the bees. Bees don't respond well to being breathed on—it can really get them buzzed off, actually.

Yet I find myself talking to them, in the same tones I use when talking to a cat that I'm holding while a vet gives it a shot. It's a tone of voice that's supposed to be soothing, but it works as well on cats as it does on bees—that is, not at all. Still, hearing that voice is soothing . . . to me, which is really why I talk.

Why do I need soothing when I work with bees and cats? Well, for one, because I know that, from their perspective, I'm betraying each of them. I'm colluding with the vet in causing pain, fear, and discomfort, no matter that it's for the cat's own good. With the bees, it's even less ambiguous. I am there to take their honey. True, I provide a house for them, but they don't know that. I protect them from predators and disease. They don't know that, either. All they really know is that I'm an invader, no better than a bear. (Well, maybe a little better than a bear because I don't destroy their hives and I don't eat their babies for tasty protein.)

So, I talk to the bees, sometimes using the soft voice that comforts screaming babies ("heyyyy, it's okay"), sometimes with the firm but good-natured voice I learned as a teacher of teens ("okay, you rowdies, settle down . . ."), sometimes with the directive voice I employed as safety coordinator in a San Francisco office building during earthquake drills ("Everybody down, I don't want you getting squished . . .").

Here's how it helps to calm me. Like that "I Whistle a Happy Tune" song from an ancient musical, it's true that sometimes the simple act of projecting fearlessness and calm is a very large step to actually becoming those things. The power of self-deception is not to be underestimated. At least, that's what I've flimflammed myself to believe.

Bum's Rush

Early last September, some fellow beekeepers reported that the females in their bee colonies had already given the bum's rush to the males, the drones, kicking them out in the cold. As I've explained, the male bees don't usually survive sex with their domineering partners, but if they do, their long-suffering sisters will make sure they don't live much longer. As winter approaches, the worker bees treat the drones like Aesop's ants treated the grasshopper—they toss the poor guys out to die in the cold.

This is an annual form of gendercide, but September is normally too early to be considered normal. The days are still warm and flowers still blooming all over the area. Speculation as to why included climate change or the bee maladies of late—mites, mysterious hive declines, and even a new one called "zombie bee" syndrome, but whatever the cause, it's probably not good news. Especially not for the drones.

Zombie Bee Syndrome???

They fly by night. *Zombees*. It's not a joke. It might be serious, yet another thing to make life more difficult for bees and beekeepers. Discovered by accident in 2012 by researchers from San Francisco State University, zombie bees have been found in three-quarters of thirty-one hives surveyed in the Bay Area, some of them belonging to fellow beekeepers who live nearby.

A parasitic fly called *Apocephalus borealis* normally infects bumblebees and wasps, but has recently made the jump to honey bees as well. What they do is not pretty: the fly deposits eggs into a bee's abdomen. As the larvae hatch, the bees become confused and zombielike, walking in endless circles by day, and leaving the hive at night, something they normally would not do. Once outside in the dark, they are attracted to the nearest street lamp, yard light, or lit window, flying obsessively against it. Their bodies can be found in the morning under light sources.

A few days later, the former larvae emerge as flies and take wing, ready to find new victims.

BEES: IT'S WHAT'S FOR DINNER?

I remember reading, when I was a mid-sized kid, that way off in the future—like maybe even as soon as 1980 or 1990, before we all emigrated to the moon and Mars in 2001—we'd be eating lots of delicious insects, including "bee burgers" made from honey bees. This was in the *Weekly Reader*, that required-reading tabloid that for many years was delivered into elementary school classrooms across the nation. I was both intrigued and grossed out about that, and the memory remains with me a half-century after.

But why not? Bees are a renewable protein source that requires no soybeans or grain, no pasture space, no special handling of odiferous waste. So, curious in the present day, I did the obligatory Google search for honey bee recipes. I wasn't surprised to get pages of hits, but 1.4 million? The top hits were for cakes and cookies. But why not? If you're going to insinuate insects (or anything) into the American diet, it's going to be through its sweets. This definitely looked like a trend worth checking out.

Except it wasn't. You probably guessed. These weren't baked goods made of bees; these were baked goods *shaped* like bees. And not even real bees, but cute cartoony bees.

Sigh. We never made it to Mars, either.

Buzzin' Cousins

Honey bees are related to a whole bunch of other insects. They are part of Hymenoptera order, one of the largest in the insect world with more than 130,000, and the suborder Apocrita, which includes all bees, ants, and wasps. All have similar bodies and all have ocelli, extra primitive eyes that augment their two main eyes by tracking gradations of light and not much else. The only difference is that bees and wasps have three; ants have only the two on either side of their heads and not the one centered in front of their main eyes.

Bees: At last count, there are 20,000 known bee species. Not all sting and only a few produce and store honey. A few of them, called vulture bees, eat dead things, but most live off plant matter. The majority of bees are solitary; those that *are* social usually gather in colonies of a few dozen; bumblebees gather in groups of 50–200 with minimal housing and no provisions for the winter. The honey bee is the only one with colonies that number in the 10,000s and that overwinter as a group.

Ants: Not all of the 20,000 ant species have huge colonies—some live in small groups of a few dozen. Like honey bees, most ants live in a large colony that consists of a queen, a handful of male drones kept for mating purposes, and tens of thousands of sterile female workers. Ants have colonized almost everywhere in the world except a few unpromising islands and Antarctica. Many people assume that termites are closely related to ants, but they're closer to cockroaches and mantids.

Wasps: A funny classification in that a wasp is defined as "any members of the suborder that are neither bees nor ants." Over 100,000 species strong, wasps are the oldest grouping of the three, from which both ants and bees evolved. Although wasps include hornets, yellow jackets, and other picnic annoyances, the bulk of them are useful in that they prey on pest insects, making them valuable for natural bio-control of crop destroyers.

Calm Before the Swarm

Hive #4 is eerily quiet. Usually, it's the active hives that attract attention, but that's nothing compared to a normally active hive that has suddenly gone silent. It's like the dog that always barks, but this time didn't.

It's lunchtime for me. I often go out to sit in the bee yard and commune with the bees while eating my sandwich. They are not wasps or yellow jackets, so they are completely uninterested in my veggie-meat sandwich or greens. (It's surprising to me how many people don't know that there's a difference between flower-sucking bees and hot dog-hungry carnivore wasps.)

I go through the possibilities. It was a strong hive, so it didn't just dwindle away to nothing in the night, and it probably wasn't killed off by a raiding party from a much stronger hive. I had seen eggs and brood in the hive a few days before, so they probably didn't "abscond"—abandon the hive as hopelessly unsuitable—because they'd never leave the kids behind except in an emergency like a fire or bear attack. Then it's probably . . . ? I'm puzzled and concerned. I notice that once in a while a single bee or two emerges and takes off. They look serious, preoccupied. Instead of first dawdling to get accustomed to the outdoors, greet friends, and soak up some warm sun, these bees act like they're on a mission of great seriousness.

And they are. Now I know what's going on. The bees I see are scouts, looking for a good location for staging what's about to follow, a swarm. About half of the hive is going to leave the hive permanently, going out to find a new place to live, taking the old queen bee with them and leaving the old hive with a new queen.

The Swarm

Hive #4 continues to give off that sort of darkening calmness, similar to what it feels like right before a gully whumping summer storm. And then it begins. Bees begin pouring out of the exits, and flying around the hive; more and more join, and the funnel cloud begins getting wider and taller, its outer edges now filling up the backyard a couple of stories high. I still sit in the middle, swiveling around to take it all in, listening to the buzzing in surround sound, feeling safe in the calm eye of the swarm.

The bee blizzard rages for a minute or two, and then starts to slowly run out of steam. Clearly, the scouts have done their job and found a safe place to gather for a day or two. Sometimes that place is hard for a human to find after the fact because once a swarm has settled in, they often hug a branch in a tight, fuzzy brown clump that, in the shadow of a tree, can look as if it's just a burl on a branch. I watch carefully where they seem to be going.

I'm pretty lucky. About nine times out of ten, each of my swarms has settled in the same place, a small tree outside my fence about eighty feet away from my hives and about twenty feet away from the sidewalk. It looks like this time is no exception. I pull myself up on my fence and confirm that the cloud has relocated loosely around the Bee Tree. Finishing my sandwich, I get my swarm-catching stuff, dress in my bee outfit, and go out there, partly to calm any passersby, but also to see if I can persuade the runaway bees into a new hive.

Within a few minutes the bees coming out of Hive #4 are acting normally now. The tension of having two queens under the same roof—a situation that either results in a swarm or the death of one of them—is abated. The ones that have sworn allegiance to the old hive with a new queen are reveling in the uncrowded living quarters, beginning to fly off in different directions, going out to do their jobs and get back to same old same old. The young drama queen is no longer stirring up trouble, and the story of a new hive continues at the Bee Tree.

The Bee Tree

At the Bee Tree, a small cluster of bees has already formed on the usual branch, about eight feet above the ground. Around them, though, tens of thousands of bees are still flying around excitedly. Looking on, I understand why people get freaked out when they see a swarm. It looks like all those cartoons of some hapless victim being chased by angry bees. But, of course, these bees aren't angry. They're stuffed with honey and they just want to find a safe place where they can chill for a while, wait, and digest. They stuff themselves because they may not be able to eat for a few days. They will wait, grouped protectively around the queen, while a few dozen scouts go looking for a permanent new home.

The scary look of a swarm actually makes sense. It tends to scare away birds and animals. Also, by moving in a chaotic mass, the bees provide tens of thousands of targets for any predator that may try to pick them out of the air, making survival of the queen that much more probable.

My tools today are a cardboard box, a sheet, a frame of wax previously used for raising brood from a hive, a stepstool, my bee suit, and a large soft brush. The brush is specially designed to gently sweep the bees off the branch without hurting them. Some old-timers use a goose wing. And the empty frame provides a smell of home and normality, calming the bees.

I am grateful for the Bee Tree because most local swarms make a beeline to it. It makes my life easier. The bees may smell the scents of previous swarms on it, because they tend to go to one or the other of two specific branches, both of them easily reachable with a stepstool. The first arrivals crawl all over it, leaving a fresh bee scent on the branch, and began fanning their wings to broadcast the pheromone message of "Yo! We're over here!" to the others.

This procedure usually works out great. Sometimes, though, something goes wrong. One morning earlier in the season, I'd watched them go through the same routine and gather on the branch. Ten minutes later, though, the swarm had shrunk to about half its previous size, and the bees were peeling off from the group in ones and twos and heading sheepishly back to their old hive. They had apparently gotten there, and the queen hadn't shown up. (I imagine her saying embarrassedly, "Oh, was that today?") They tried again, more successfully, later in the day.

Without my intervention, they'd typically stay on this branch for one to two days before scouts found some suitable place and led the cloud of bees to it. I don't want them to do that for a couple of reasons. For one, I'd rather have them be a part of my apiary than try to establish a new home somewhere else. For another, this isn't a wild place, so they'd probably find a hole in the wall of a house or someplace similarly likely to get them exterminated. So, my job today is to convince them they'd be happier in an empty hive body and back in my apiary.

It isn't so hard. If you provide good surroundings, they'll usually stay. I lay my old bedsheet under the swarm, settle the comb firmly inside the cardboard box to give them something to hold on to, and climb a step or two up the stepladder.

The bees are peacefully clumped together now. I hold the box a few inches under the mass of bees, grab the branch they're resting on, and give it a sudden, solid shake. Since only a few hundred bees are actually gripping the branch and the rest are holding on to them, a good sudden shake will result in about 80 percent of the bees falling into the box. Most of the rest can be gently swept into it with the bee brush.

Everything works the way it's supposed to. Mostly. There are always the hardcore holdouts that won't leave the tree, the bees that end up on the ground below, and those that fly instead of falling into the box. Frankly, though, there's only one bee I especially hope ends up in the box: the queen. If she's in there and doesn't flee, the rest of the bees will eventually follow her. To this end, I lay the box on its side on the sheet and mostly close it up, leaving a box flap partly open.

The queen is in there, and I see bees fanning her scent at the openings of the box and straggler bees landing and crawling in. If she weren't in there, bees would be crawling out and returning to their place on the branch . . . or wherever she decided to fly next. I leave the box there for an hour in the shade so the scout bees can find their swarm-mates when they return to find the branch empty.

Later, I gently shake and brush the bees into an empty hive body and watch to see that they start fanning their scent from the hive entrance. They do. The queen's inside, all is well, and they're broadcasting to any latecomers, "We're over here! Welcome home!"

The Buzz in the Marketplace

Even in winter, bees make a slight buzzing sound. Flapping their wings slightly is a way of generating heat. Most insects, as you might recall, are cold-blooded—they don't generate heat; they use whatever is provided to them. If it's warm, they get active; if it's cool they get sluggish (as do, coincidentally, slugs). If it's cold they stop moving completely and act dead. (Sometimes the act is so convincing that they stay that way forever.) Bees, in contrast, are able to generate minute amounts of heat by aggressively flexing their wing muscles like buff guys waiting to use the weight room. Not enough to keep themselves alive if they were alone outside, but enough to keep themselves, their queen, and their grub-like brood alive when crowded together by the thousands.

In the summer, they use their wings in a different way, this time to keep cool. Approaching the colonies in the heat of a summer afternoon, a beehive resembles a modern office building. If you've ever gone into a business district early in the morning on a weekend, before the noise of traffic starts in earnest, you may have noticed how loud buildings are. Their ventilation systems roar in a loud hum, moving a vast amount of air through a building where the windows don't open. Well, beehives are the same, only in miniature.

In a quiet bee yard far away from traffic, I can hear a hive's ventilation system from a distance of thirty feet, that powerful, buzzing rush as lines of bees pump air through their sun-heated, windowless home using nothing more than hundreds of their tiny wings.

Listening to the hum, watching the bees bustling in and out the doors, sometimes flashes me back to my corporate days working in a subgroup of a subdivision of a subsidiary of a major phone company. I can imagine the bees in business suits as they enter the hive, flashing their badges at the guard bees, texting and talking on their cellphones on the way to their cell-like cubicles. Each with a plan for the day, a schedule, an assigned set of tasks, and a grim determination to get them done.

A bee produces a teaspoon or two of honey as its life accomplishment.

A bee produces a teaspoon or two of honey as its life accomplishment. When no longer able to do a full load, the hive thanks her for all the hard work by tossing her out of the hive to die. And no, I'm not presenting this as a metaphor for our times.

> NUTS TO ALMONDS!
>
> Every February 1.6 million beehives converge from all over to pollinate California almond fields. The problem isn't just finding that many healthy hives in February, before the honey season has really started; it's also that the bees really don't like the pink and white almond flowers and will avoid them if anything else is available. Almond honey is unusual in that it's very bitter, drowning out honey's normal sweetness with a retchy medicinal taste. (I was once given a spoonful to taste—one of those practical jokes beekeepers find hilarious.)

Graduation Day

If you go out into the bee yard at the right time of day, you will suddenly witness hundreds of bees tumbling out of each hive, frolicking around the entrance before flying exuberantly around their home. These are the graduates, the new bee "newbies" leaving the hive for the first time.

A week or two earlier, the new bees had emerged from their cells and immediately gotten to work. Their first job had been cleaning out the cell they just hatched out of, but then they had taken on new in-hive duties, such as cleaning, building comb, guiding the queen to empty cells ready for eggs, taking care of the brood, guarding the hive from interlopers, relocating pollen and nectar, etc. Finally, though, they became ready for this day: graduation to a new job, in an infinitely big world, as field bees, collecting nectar and pollen.

They start out cautiously, hugging the outside of the hive, facing the entrance as if worshipping it, using it as a familiar sight and smell of home as they adjust to the brightness of the sun and strangeness of the outside air. After a while, they start flying slow, tight little sideways figure eights, the math symbol for infinity. This is the first time they've been airborne, and they test their wings while keeping the hive entrance always in sight.

As they gain confidence in their abilities, the bees begin swooping and banking, intoxicated by the world's vast sensations and sights, by the smells of blossoms wafting in on the wind. Their circling becomes wider and more freeform. Yet they keep making figure eight patterns with their hive at the center of the crossover. You can see they're balancing between exploring the world and reassuring themselves that they know where their home is.

Finally, having burned the details of their home location into their memories from all directions and angles, they're ready to go. With one final swoop they take off in a beeline, heading straight to who knows where, following a faint scent that promises distant blooms of unknown sweetness.

Watching this always brings a sense of déjà vu, of the eternal figure eight dance every person, perhaps every animate thing, does from the moment they can move independently:

giddy loops of exploration balanced with needing repeated assurance that there's a way to get back home again.

Watching my bees, I feel like a parent on graduation day, looking on with pride, no little worry, and a realization of my own powerlessness and irrelevance in the process. The onlooking beekeeper can only hope they'll find what they're looking for, not get blown off course or eaten by a bird along the way, and return safely again.

WHY BEES MAKE THE BEST PETS, TAKE 2

1. Bees don't bark and whine all night if you leave them in the backyard. In fact, they rather prefer it.
2. Bees don't crowd you out of your bed at night.
3. Two legs good, four legs better, six legs best of all.
4. Bees aren't needy. They don't demand petting, attention, or a food dish. They find their own food.
5. While cats and dogs may be generous with their gifts, headless mice and dog poop are not anything you really want. Bees greet you with honey for your toast and beeswax for your candles.
6. If, God forbid, you die without anybody knowing, your hungry cat will start chewing on your body within a day or two. Your hungry dog will hold off for several more days. However, no matter how long you may lie there, your bees will ignore you and go about their business.

7. You will never be tempted to succumb to your worst self, dress your bees in funny costumes, and humiliate them on YouTube.
8. When bees pay attention to your plants, they don't dig them up or kill them with pee. They actually help them blossom, bear fruit, and thrive.
9. You don't have to worry about keeping them locked up in your yard. They don't stay there, of course, but you don't have to chase them around the neighborhood—they come back without prompting.
10. Hundreds of individual bees are born each day, and hundreds die. You learn a lesson about life, and will never be tempted to spend $3,000 on some expensive procedure at the vet's office.

Uncooked Honey, Raw, Raw, Raw!

If you tell me you believe in the healthful benefits of using raw honey, that's great. But please don't add that you take a dose every morning in your morning tea. Raw, you might well know, just means that the honey hasn't been heated above a certain temperature. So, of course, putting it in tea or baking it in a recipe immediately destroys its "raw" status and any benefits thereof.

I have long resisted putting a "raw" claim on my honey labels for reasons both rational and irrational. For one, I've assumed that it's generally understood that no small reputable beekeeper would provide honey in any other state but how it comes out of the hive.

Another rational reason I dislike the term is that "raw" is legally meaningless. Some honey sellers define it as never being heated above 90 degrees; some stretch that up to 110 degrees; and some realize that they aren't legally prevented from making the "raw" claim no matter what they do.

You can be assured that large-scale supermarket honey isn't raw and has probably been heated up to about 150 degrees. Why would they do that? It isn't because honey needs pasteurization—because of its low moisture, honey doesn't grow bad organisms. (Or *good* ones either—in order to get it to ferment for wine or beer, you have to add water to it.)

No, industrial honey packers heat honey in order to make it thinner and easier to work with. Honey at room temperature is a pain. It requires more effort and infinite patience as the thick liquid *sllooowwwllly* drips from one place to another. By heating it, the mass honey factories can pump heated honey out of the combs, down long tubes, through filters, and into jars like water in a few seconds flat. Time is honey!

Honey at room temperature is a pain.

When I extract honey, it takes more time and more muscle with my hand-cranked centrifuge, but it's worth it. Besides the sometimes-claimed health benefits, honey's fruit and floral flavor notes are easily evaporated away with heat, quickly turning a sublime taste adventure into bland industrial supermarket sugar water.

Still, although claiming "Raw!" should be a marketing advantage, I've resisted for reasons I couldn't quite put my finger on. Finally, I realized why; it's the word. In almost everything *but* honey, "raw" carries a lot of negative baggage. It implies things like crassness, roughness, immaturity, unreadiness, even danger: Raw meat. Raw deal. Raw data. Raw sensibilities. Raw skin. Raw emotions. Raw language. Raw sex. I fear that for every health food aficionado who would be attracted by the term, others may react, as I did, with an unconscious aversion. So my label still doesn't have the word "raw" on it.

Mead Eaters

How do bees tell when the nectar has dried into honey and is ready to seal up with a thin layer of wax? The answer is, I'm not sure anybody really knows. Probably not by time, but maybe by thickness, since different humidity, heat, and even nectars will dry into honey at differing rates. We do know that it's important to get it right—watery honey will ferment, making a mead that can kill honey bees, even though it's something humans have made on purpose for eons. Which only goes to show that "one man's mead can be a honey bee's poison."

Meading with Friends

I'm lucky that I live within a few miles of several urban wineries and even a couple of distilleries. Years ago, a friend of a friend introduced me to the Smith brothers. No, not the bearded cough drop makers from my childhood, but Matt, who is a professional winemaker, and Dave, a professional distiller. They wanted to make mead and knew that I had a lot of honey.

Why not? I thought. Back then, I thought mead was a super-sweet dessert drink (and it is possible to make something like that by killing all the yeasts that normally turn all available sugars into alcohol). Instead, Matt and Dave were going for authentic mead. That takes a long time to ferment and an even longer time to age. Luckily, they knew what they were doing.

Months after mixing up the ingredients, our concoction emerged from the barrels. Frankly, it didn't taste that great at the time. Dry. Sort of like mediocre white wine. Still it was nice to put it into bottles and cork it in a professional way, not fully believing it would become any better with time.

I pulled out a bottle every three to six months and tasted it, and it did smooth out over time, turning from a bad white wine to a pretty okay white wine. That was maybe five years ago, and I think I still have a bottle or two somewhere. I may open the last one sometime, but for now, I'm not terribly fond of white wine anyway. However, the Smith brothers and their families have become treasured friends, and that is as sweet as the mead is tart.

Feeling Down?

The way to tell the age of a honey bee is by whether it still has a lot of golden baby down on its thorax. Older bees, constantly squeezing past other bees in the narrow aisles in the hive, will have worn away much of their golden fine fuzz. Younger bees haven't, and feel like a kitten or a baby chick. Sometimes if I find a calm young bee, I take off my glove and gently pet it. Soft!

> Bees, unlike most other bugs, generate their own heat instead of merely adapting to the outside temperature. They keep the center of the hive a toasty 34 degrees Celsius (94–96 degrees Fahrenheit), whether winter or summer.

Hexagon, But Not Begotten

One of the mysteries of bees is how they create their combs with perfect six-sided cells. It's pretty amazing, but it's not as mystical as it seems. The method is actually practical and commonsensical. The key is that the cells are not really hexagons. They're round. They are the size that a bee can fit comfortably into, because the bees working from inside the cells use themselves as a measuring device. Why do they seem to be six-sided? Easy. Try this. Raid your penny jar and put a penny on the table. Now arrange coins in a circle around the outside of the first coin. How many will fit? Six. So, simply building round combs as tightly as possible gives the bees something very closely resembling a hexagon.

An Audience with the Queen

I only very occasionally see the queen inside the hive. That's intentional. I try to avoid bothering her and the babies—the eggs and larvae that require a lot of care and warmth to stay alive and develop right. As a result, the bottom couple of hive boxes are more or less her private chambers, as far as I'm concerned, only to be invaded if it looks like there's a serious problem, which I can usually tell just by observing the workers' behavior outside the hive. A hive with a good queen has lots of bees happily flying around it. It especially has a good proportion of bees arriving fully laden with pollen. Pollen is full of proteins and is in high demand by the bee larvae.

A queenless hive will show it, eventually by dwindling down to nothing, but initially by moping. If you've ever had an ant farm, you know what I'm talking about; without a queen, listless bees begin hanging around the entrance, not doing much of anything, looking like sullen, bored teenagers. That is, until I decide to get closer to take a good look—then the sullenness becomes belligerence, with bees seriously trying to sting any body part they can get to.

Because a healthy queen is important, some beekeepers make a point of going in and looking for her, inspecting her handiwork to see if she's laying lots of eggs, and replacing her at the first sign of trouble. However, looking for the queen is a lot like playing "Where's Waldo?" You can usually find her by looking carefully in the center of the hive where the eggs and brood are (not the edges and top where the honey is). It helps to look at arm's length for a half dozen bees in a "daisy" pattern—a circle with their heads all pointing toward the center of it. Look for a larger bee inside the circle that's got a long pointy rear end (the better for backing into a cell to lay an egg in it a thousand times a day). That's the queen.

Those workers in the "daisy" pattern are the queen's ladies in waiting. They groom her and feed her, and lead her to the next piece of empty comb in which to lay eggs. In most ways, the hive's "queen" is really its captive. Her attendants supervise where she is and what she's doing, twenty-four hours a day. They constantly evaluate her as well. If she isn't performing the job well, whether because of illness, injury, or age, the signal goes out through the nursery: "Revolution! Start growing a new queen." (More about this on page 8, "The Sex Workers.")

> Some beekeepers "cheat" by painting a small dot of paint on the back of the queen's thorax, color-coded by year so they can not only find her easily, but know her age as well. For example, the bee industry standard is that a queen born in 2011 has a light gray spot; 2012, yellow; 2013, red; 2014, green; and 2015, blue, with this color sequence repeating every five years (since queens rarely survive more than a few years).

Busy as a Bee?

By their reputation, you'd think all honey bees would be busy all the time.

Not necessarily true. Each colony seems to have its own culture. Some "Type A" hives have bees coming and going from sunrise to sunset. Other hives, however, are frustratingly "Type B." Their landing board at 11 a.m. looks like your local hipster coffee shop, bees standing around in the sun, talking, laughing, and sipping decaf espresso. It sometimes makes me want to blurt out, "Don't any of you have jobs?"

And yet . . .

When I compare actual productivity of the two types, the results are surprising. Their honey production is about equal. It turns out that the Type B bees *do* work. They get going by noon and work at a frenetic pace during the heat of the day when most flowers are producing their highest levels of nectar, knocking off long before dinner time.

The bees' landing board at 11 a.m. looks like your local hipster coffee shop, bees standing around in the sun, talking, laughing, and sipping decaf espresso.

Do Type A bees just pace themselves during the day so they work longer but easier? Or are they really working harder, but neutralizing the time advantage by (for example) eating up more honey in traveling farther and visiting more blossoms in the less-productive morning and evening?

Regardless, as a Type B type worker myself, I take comfort in this example, and given the choice, would rather live in a café culture colony. (Hey, drone, over here—another latte, extra honey.)

MEDITATION ON A BEE STING

The good news is that I have 499,999 bees that *didn't* decide to sting me on the finger today.

Clipped Wings and Prisoner Bees

Sometimes a colony of bees will want to relocate to a better place, or split a crowded hive in two. Some beekeepers try to prevent this by clipping a wing of a queen. If she can't fly, their thinking goes, the bees will have to stay even if they are unhappy.

This is cruel on several levels. Ironically, it is also self-defeating for the beekeeper. Unhappy prisoner bees will not be producing much honey. Worse, the beekeeper loses huge educational opportunities, because if you can learn how to keep free bees content while still getting what you want in return, you've gained great lessons in beekeeping and—dare I say it?—life.

Stationary Flying Is Very Cool

Bees grip the edges of the hive entrance and buzz their wings, "flying" while standing still. What gives? They are part of a long line of bees all facing the same direction and fanning a breeze through the honeycombs, drying the watery flower nectar into thick, sweet honey.

FOUR SEASONS OF HONEY

The annual rhythms for beekeeping parallel those of gardening. That makes sense considering the symbiotic relationship of bee to flower. As with anything that you do outside, every season provides different variations and challenges, as well as preparations for what's coming.

Winter: Busy as a Beekeeper

Winter is unusual in that it's the only season when the beekeeper works harder than the bees. While the bees' job is pretty much to survive, the beekeeper has to build, repair, make plans. . . .

continued

For an aspiring beekeeper, winter is the best time to get started. To begin with, there's a certain level of construction necessary to put together a beehive. Some are willing to pay a premium to get everything delivered prebuilt, but for most of us, getting the hive bodies together means buying precut kits of the boxes, frames, and beeswax to assemble and paint. (While it's possible to build hives from scratch using plans and lumber, nobody would even think of it, unless they have a lot of power tools and an inordinate love of woodworking.)

Besides building new hives, there's also maintenance on the old ones. Before winter I took off the one to four "upper stories" of the hive (called "supers," where bees store the excess honey), leaving the two stories of "hive bodies" (where bees cluster together to keep warm with the brood), and enough pollen and honey to keep the colony alive through the winter.

This pile of empty supers is ripe for inspecting, repainting, repairing, and/or replacing. There are things to watch for: wax worms, for example, the larvae of a moth that somehow manages to find far-flung and remote beehives and lay eggs inside them. An active colony will discover the larvae and eject them, but take the boxes off the active hive and there's nothing to prevent them from chewing up the precious wax combs and leaving a mess.

The winter is also when you need to figure out if you'll need new bees and decide how you're going to get them. There are several choices. Established beekeepers may decide to split a strong hive into two. They may decide to wait until swarm season in early spring, with the hope of catching a few swarms and placing them in their empty hives. For many new beekeepers that sounds daunting. They (or experienced ones who want a specific type of bee) will reserve package bees to be shipped or picked up in the spring. But that is also a winter task. Usually the bees are sold first come, first served. If you don't reserve bees early in the winter, you might not get yours until weeks or even months into the spring, decreasing the chances of getting a significant amount of honey the first year.

continued

Winter is also a good time for reading bee guides, brushing up on beekeeping knowledge and skills, and looking for new ideas or answers to questions you might've had.

There is, of course, no honey. In fact, much of the time, there's no visual evidence that there are any bees in those hives. As with gardeners and their seed catalogs, the catalog from Dadant's Honey Bee Supply Co. serves as a first sign of midwinter—an encouragement and comfort to the impatient, winterphobic beekeeper.

During any sunny winter "warm spell," the bees make a brief cameo, taking the opportunity to fly out and purge long-filled digestive systems, returning to the warmth of the hive as quickly as possible. I wonder if I should let them use my iPod during the rainy days ahead. I figure I can load it with the Honeys, the Honeycombs, the Buzzcocks, B. Bumble & the Stingers, the Bee Gees, and the Bird and the Bee. Perhaps even an assortment of B-sides and B movies like *Bee Season* and *The Sting*.

Spring

The bees are impatient about the late season rains. You can see them staring wistfully out of the hive entrances like school kids missing recess. A break from the rain gives them a chance to raid the flowers blooming all around us, and the intrepid beekeeper a chance to check out the progress so far. They keep promising that the sweet stuff's coming soon, but he replies, "Show me the honey!"

When we harvest honey, the smell of sweetness and beeswax fills the house. The bees are very busy and the spring honey comes in with wings on. If you've never had fresh unprocessed honey before, it's amazing how many different flavors you can taste in there besides just what you'd identify as the dominant "honey" flavor. It's like the difference between a freshly picked garden veggie and a store-bought one, or a really good wine and a mediocre one. The honey going into jars is

continued

light-colored and light-flavored, as bright as the spring flowers it comes from; as the summer and fall progress, the honey will get darker and the flavors will become even more complex.

Summer

The summer honey is flavorful and florid, the color darker, as we move out of the subtle spring flavors into the more robust summer ones.

Autumn

The last few harvests of honey in the fall have a slightly sad feel to them,. Another harvest season, another ending of a productive year. The honey is interesting, though: dark, with a molasses flavor to it and a slight grace note of eucalyptus, tasting a little like a minor key chord, a bit of autumnal melancholy, of golden sunsets, lengthening shadows, and cool autumn nights.

Interspecies Understanding

When I had dogs and chickens, I let them wander freely near the beehives. There are two schools of thought about bees and other animals (including kids above a certain age):

1. Protectively fence off a small portion of your yard, giving six to eight feet of clearance at the front of the hives. I don't like this solution. While it protects the animals, it makes that much of the yard unusable. If you have chickens, they'll be deprived of the tasty dead and dying bees, and you'll have to keep trimmed the grasses that the chickens normally keep short between hives—a place where you may not want to go with a mower.

2. Let the animals and bees work it out—bees are usually pretty tolerant, unless they feel directly threatened. If, for example, a chicken decides to go up to the front of the hive and pick off returning bees, a guard bee will give a threatening buzz and stingless dive-bomb that most other organisms retreat hurriedly from; if not, it usually doesn't take more than one sting to train any animal or fowl. Or person, for that matter.

Spring Forward, Little Bees? No, Fall Back!

It was an unusually warm winter day, a week before the winter solstice, in a place where the climate is pretty mild. When I say warm, I mean as in the lower 60s with no appreciable wind. I missed my bees, now consolidated into just two strong hives to make it through the winter rains, cold, and darkness, so I ambled into the b'yard—my backyard as well as my "bee yard."

Both hives were located right next to the southern fence. The sun at the solstice is so low on the horizon that the hives weren't getting direct light during the day. I thought about moving them into a warmer spot, but doing so is a several day project of incremental moves. That's because bees get so oriented to an exact hive location that they'll get confused if their home moves more than maybe six to twelve inches at a time. They'll reportedly go to the spot where their hive had been, and just stand around until it gets dark and they freeze to death, alone and confused.

(At least, that's the story often told in beekeeping circles, although you'd think they'd be able to figure it out by seeing or smelling out their hive a few feet away. But in the winter, I really didn't feel like testing that theory out on *my* bees, thank you very much.)

Anyway, despite being in perpetual shade, they were flying off in all directions. No lolligagging around the front of the hive, more like a fast-motion movie of JFK Airport. *Vroom! Zoom! Whoosh!*

I slid around the side of their flight path and sat on an empty hive a few feet away to watch. I was dressed in light colors, out of the sight lines of any guard bees that might be lurking around inside the front entrance, so I (correctly) guessed that they'd ignore me.

Sitting there was good for my soul, like spending time with an old friend I hadn't seen for a few months. They looked healthy. I watched the incoming flights, looking for pollen hanging off their legs. I didn't see any, which I took as a good sign. Pollen is for the larvae. This time of year, the bees shouldn't be worrying about that; they should be minding their honey supply to make it through the weeks when they can't get outside. It's a few months too early to be raising young ones for the new season, but sometimes they get confused by mild weather, potentially setting themselves up for too little honey and cold-killed larvae when they get hit with winds and 38-degree rains in January and February.

Luckily, something's usually blooming around here year-round. The bees might find some late clover blossoms or some early eucalyptus blossoms, both good for honey; finally, in very early spring, at the time when they'll start thinking about ramping up on the raising of babies, they'll find the feral rosemary that blooms in the winter next to the shoreline. Forget about robins or groundhogs or returning geese; the harbinger of spring for me is seeing the rosemary pollen, white as the coastal fog, coming in on little bee legs.

No Snow

Yes, I am grateful that I'm not raising bees in a place that has real winters. It's possible to do it. You may have to wrap up your hives in insulation, leave a lot more honey inside for the bees, and brush the snow off the landing boards during every storm so they won't suffocate, but it's possible. I'm just glad I don't have to do any of those things.

SMELLS LIKE BEE SPIRIT TO ME

In the winter I burn a candle when I write. It focuses me, and keeps me from wandering away from the desk. Not just because I'm somewhat phototactic like a moth, drawn toward a flickering light source, but also kept there, being somewhat afraid of leaving a fire untended, presumably like a ladybug afraid her children may burn.

That's only part of the reason I do it, of course. I burn beeswax candles from my own hives. I love the smell—who doesn't? It's such a mix of the smell of flowers' nectar and the bugs that eat it exclusively. If there were such a thing as flower fairies, this is what they'd smell like—equal

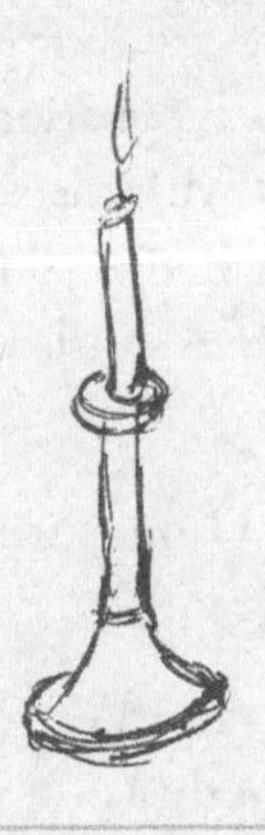

continued

parts animal and vegetable with a little magic binding it together in a pleasing balance.

The inside of a healthy beehive smells something like this, but it's an indescribable thing as well. The easiest way to explain the aroma would be to simply send you on a field trip to a bee yard, to have you standing close when the beekeeper opened the hive, but that's an assignment I can't facilitate or enforce, so let me try to explain the scent. Imagine honey. It's got a purely floral smell, right? Now imagine the smell of a beeswax candle. A little different, right?

Now, imagine subtracting the smell of honey from beeswax. What you have left is the smell of the bees. Multiply that leftover part of the smell by two. Wait, you're not done yet! Okay, now the hard part of the math equation. Imagine that doubled wax smell, but with the smell of the honey added back. That will give you approximately what the inside of a beehive smells like. If it would help, here's the equation:

(W[ax scent] - H[oney scent]) X 2 + H = B[eehive scent].

Now, isn't that clear? What, no? Well, don't feel bad, not everybody can be mathematical.

Bee Beards

If you want a beard of bees, you have to be willing to be stung a few times. That's if things go well. If they don't go well . . .

You've probably seen pictures of bee beards, sported by brave and foolhardy people who have arranged to have the lower half of their faces covered with stinging insects, as if they're auditioning for some kind of entomologist-only ZZ Top tribute band. It's an old pastime among thrill-seeking bee fanciers, because it impresses the rubes.

Ukrainian beekeeper Petro Prokopovych modeled the first known bee beard in the 1830s. Besides the bee beard, Prokopovych was also the inventor of several other innovations in beekeeping, including removable frames and the first queen excluder (a slotted board big enough for workers but too small for the queen to pass through, keeping bee eggs and larvae out of the honeycombs). Demonstrating what he'd learned about bee swarm behavior, he placed a captive queen in a cage under his chin and released thousands of bees near his face. Sure enough, the bees went into the typical swarming behavior of bunching tightly around the queen, creating a "beard" that hung off his chin.

His bee beard inspired imitators. Carnival freak shows decided that their discerning audience needed to see *this*, and bee bearders became as popular as fat ladies, Siamese twins, dog-faced boys, pinheads, and the Wild Men of Borneo.

The thing is, a bee beard isn't hard to accomplish, but it takes some guts and a willingness to be stung a few times. This is not recommended for children, or anybody with an allergy or aversion to bee stings, any level of good sense, or a litigious lawyer. Please don't try this at home . . . or anywhere else. That said, here's what you'd need to do if you were inclined to create a bee beard. (This is from experts; you don't think I'd be foolish enough to actually do this, do you?)

- Bee bearders select a hive with easygoing bees willing to put up with their outrageous shenanigans without exacting too much revenge.
- They find the queen and lock her in a queen cage (a small wooden box with metal screening on one side that sort of looks like a kazoo that someone made in woodshop).
- For a lush, full beard, they need about 12,000 bees (two pounds). Bee bearders box them up with the captive queen the day before, keep them in the dark, and feed them well. Spritzing them with sugar water is said to work pretty well for this. The hope is to calm them.

- When they're ready for the beard, the bearders tie the queen cage under their chin. protecting their eyes by wearing swim goggles. Bees will be crawling everywhere, so they cover their hair, button their shirt top, tuck their pants into their socks, put cotton loosely in their nostrils and ears, and put some petroleum jelly around their mouth and eyes with the hope of maybe discouraging the bees from crawling into either.
- Remaining calm from this point on may seem counterintuitive, but it is very, very important. They open the box of bees and hold it against their chest so the bees can smell the queen. The bees, in the best case, will peacefully begin crawling up their neck to surround the queen's cage, hanging in bee garlands from the bee bearder's skin and each other.
- If all goes well, the worst that will happen is that the bearder will have to get used to the sensation of thousands of bees gripping their skin with barbed bee feet.

- The bee bearder can pose triumphantly for photos. Then, while things are still going well, they have their assistant untie the queen cage and place it back inside the box into which they want the bees to return.
- One bee bearder trick is to stand over the box, jump up into the air, and land hard. This dislodges the bulk of the bees onto their feet and the ground. The befuddled bees will again smell the queen and begin crawling into the box. Eventually, all of them will be back in captivity, ready to be transported home again.
- One last note: Bee bearders say they hope for the best but prepare for the worst. Odds are pretty good that they'll get at least a sting or two, even if they do it right. However, even experienced bearders have sometimes misjudged the bees, the weather, the cosmic influences, or their own calmness, and been stung dozens of times. They have to be prepared for the medical ramifications of that and the potential of bees attacking civilians. Although they dread having to use it, pros keep an emergency sprayer filled with soapy water that can kill masses of bees and an industrial shop vacuum to dispose of the evidence.

Extreme Bearding

As if a two-pound beard isn't impressive enough, there's been a trend toward "beards" that cover the whole body as bee bearding became a competitive endeavor. The world record of 87 pounds of bees (approximately 350,000 of them) was set in 1998 by American Mark Biancaniello and still stands. However, if you're judging by style points, give it up for a couple of beekeepers from Ning'an, China, Yan Hongxia and Li Wenhua, who were married while covered in bees. It is a wedding fad that has yet to catch on. But if the bride's wearing a veil anyway, and you want a little extra honey in your honeymoon, why not?

Apathy for Apitherapy

Here's what I'd like to believe about bees: They are in cahoots with a Higher Power to do what's right for the Benefit of Humanity and that all of their products are good, healthy, and therapeutic, that they're so therapeutic that the Medical Establishment has tried to suppress the Truth.

Oh, I would like to believe that on so many levels. First of all, I'd love to be the holder of esoteric substances and secrets that *they* the powerful don't want us to know. I'd love to believe that there are forces of Good and Evil in the world and that my bees and I are squarely on the Good side along with angels, kittens, chocolate, and Bruce Springsteen. Next, I'd love to believe (despite only scant and anecdotal evidence) that a benevolent God and the bees love us and want to help us.

I'd love to believe that there are forces of Good and Evil in the world and that my bees and I are squarely on the Good side along with angels, kittens, chocolate, and Bruce Springsteen.

Friends and acquaintances love telling me about new cures and therapies somebody's trying from the Medicine Chest of the Beehive. Of allergies prevented, lingering wounds healed, and so on. I just say "Hmm!" and make a note to look into it. I am genuinely glad if somebody is suffering less than they were. However, I remain a great believer in the genuine powers of the placebo effect, and I am less sure, in fact skeptical, about most of the claims of "apitherapists" (people making claims that bee products are powerful medicines). Let's take a look into the apitherapy medicine chest:

Propolis

Sap is a great thing for trees. It acts like blood, transporting nutrients from roots to top branches. Like blood, it also clots when the tree suffers an injury, sealing the wound with a sticky spot that can protect it from microbes and bugs.

Tree sap is also a great thing for bees. They collect it and use it to fill small gaps in their hives, keeping out cold, bugs, and microbes. Taking the idea that if it's good for beehives, it's also good for you and me, alternative medicine companies sell propolis pills, tinctures, toothpastes, and more in health food stores.

There's clearly more mystique in selling propolis—selected, gathered, mixed, and trampled by bees—than merely selling random "tree sap." That's okay. If it works, great. And it would be comforting to believe that, as six-legged shamans, bees are especially astute at selecting the very best, the most healthy, the most wonderful tree sap for their propolis.

If only it were true. In reality, they select pretty indiscriminately, looking for a certain workable consistency over other concerns, which means that propolis in a hive can also include tar, drying paint, petroleum jelly, that Tanglefoot sticky stuff people put around tree trunks to keep ants away, and other not-so-healthful stuff.

Still, on the other hand, if it's pure tree sap—and how can you tell, either way?—odds are good that it won't do any harm. Especially so if you happen to be a string instrument. Antonio Stradivari, the guy who made Stradivarious stringed instruments (but not the Stratocaster electric guitar), reportedly used propolis in his varnish to accentuate the wood grains of his violins. Look for it in not just health food concoctions but in some chewing gums and car waxes, too.

Royally Screwed

Royal jelly is in various health and beauty products. It has also shown up in some alternative medicine discussions as being a fertility aid for women trying to get pregnant. I guess this is the thinking:

Royal jelly is what worker bees fill queen cups with. A queen cup, of course, is an oversized comb cell that a lucky random larva gets plopped into, sort of like a random toddler being chosen from all others to become the Dalai Lama. The extra royal jelly gives the random larva working ovaries, making her a queen.

What is it? It's a secretion of the worker bees. It comes from their hypopharynx, a globular structure in their upper throat.

Royal jelly is what worker bees fill queen cups with.

While it's true that a queen gets a generous portion of royal jelly, the worker bees get it, too . . . just not as much. (Actually, the drones get it too, so it's not just a gal thing.)

To get it from the bees requires stimulating the workers to create as many queen cells as possible, waiting for the bees to fill them with royal jelly and a grub, and drawing off the liquid by hand with a hypodermic needle or tiny vacuum. From each hive, it's possible to get about a pound of royal jelly—something less than a pint—during a season of aborting future queens.

Does it do any good for humans? Well, probably. It's got some proteins, amino acids, and B vitamins (or, if you prefer, *bee* vitamins). But it can also do some harm as well; if you're allergic to bee stings, there's a good chance that you'll be allergic to royal jelly as well. Will using it get you pregnant? I dunno. Probably no more (and no less) than eating at a certain pizzeria or standing on your head after sex—the placebo effect is a pretty powerful thing. Maybe there's something to it, but it almost makes you wonder if some people take that whole "birds and bees" sex talk a little too literally.

THERAPEUTIC BEE STINGS

You gotta be kidding, right? Just because something hurts doesn't mean it's good for you. There's no scientific confirmation of the anecdotal reports that getting stung on purpose (for allergies, arthritis, or the symptoms of multiple sclerosis) has any beneficial effects.

Ethel, the Indoor Bee

When bee pheromones were first discovered, many researchers assumed that they were used only by the queen to issue orders to the workers. However, it turns out that lots of bees in the colony emit chemical signals for all sorts of things. For example, ethyl oleate is a pheromone that inhibits young bees from becoming foragers. It's a kind of alcohol that forms in the honey-storage crop of foragers and its vapor is released when foragers release nectar to the young housekeeping bees. So, if there are enough foragers in the colony, the ethyl oleate they release keeps the young bees working inside the hive. However, when there are too few foragers, the shortage of ethyl oleate stimulates some of the young ones into becoming foragers.

Bees also release pheromones that stir up the defenders of the hive when defending against an intruder.

WHY BEES MAKE THE BEST PETS, TAKE 3

1. Bees are cool.
2. You don't have to buy a pit bull or a Doberman to have people think you're brave and tough.
3. Bees have a quiet dignity, except when they're trying to sting you.
4. Then, they have righteous indignation.
5. You can't make a living beard out of birds, lizards, snakes, or mammals. Not even a chin warmer from a chinchilla.
6. Bees live in a matriarchal society. The women run everything and the males are boy toys, desired for sex only. (Isn't that pretty much what most human men and women secretly fantasize about as well?)
7. Bees make good role models. They're hardworking, fiercely loyal to family, and slow to anger (but not pushovers).
8. Bees house-train themselves.
9. Bees don't track mud, poison ivy, or fleas into your house.
10. Bees don't have kittens.

Got 'Em? Smoke 'Em

Eons ago, some cave dweller stumbled upon a new use for their relatively new invention of fire. It turns out that the smoke from it has a strange calming effect on bees when you're trying to steal their honey. Instead of trying to sting everything in sight, they crawl into their hives and start gorging themselves with honey.

Beekeepers still smoke their bees. They use a smoker, which is sort of a steampunk metal thing, roughly the size of a half gallon of milk, looking sort of like the Tin Man's oil can in *The Wizard of Oz* with a bellows attached to it from a tiny accordion sailors sometimes play in old movies. You light a fire inside from the top, which is hard to get going without getting your knuckle hairs singed, and add some reasonably natural fuel—dried paper pulp, pine needles, dry leaves, scraps of 100 percent cotton T-shirts—that will smoke nicely. The smoke emerges from a spout on the top when you close the lid and gently squeeze the bellows, which are small enough for one-handed operation.

The reason smokers look so delightfully old-fashioned is that they are. Perfected by beekeeper T.F. Bingham of Farwell, Michigan in 1875, they really haven't changed much at all.

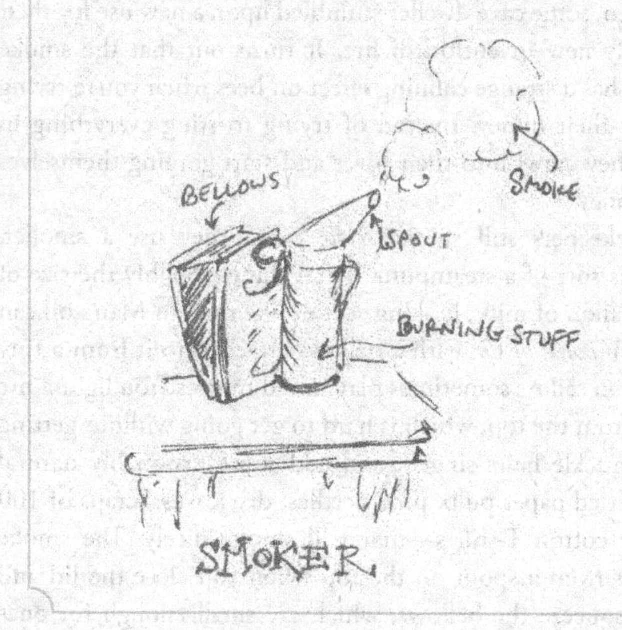

Why does smoke have that calming effect on bees? Three reasons. First, they are triggered to go into fire-drill mode, getting prepared to abandon the hive if necessary, which means they're distracted with stuffing themselves with food. Secondly, once they're stuffed with honey, it's hard to bend their abdomens enough to sting you even if they wanted to. Finally, the smoke masks their smell of fear and panic. Bees release alarm pheromones when under attack, but smoke masks the smell, short-circuiting any concerted defense of the hive. (Bees also leave a telltale chemical scent where they sting, creating a big bull's eye for other bees to aim at, which is why smart beekeepers quickly smoke any place a bee stings.)

Bees also leave a telltale chemical scent where they sting, creating a big bull's eye for other bees to aim at . . .

Assassination of a Queen

I hate killing bees by accident, but it happens. There are so many of them running around unpredictably and there are so many heavy parts of a hive that it's almost inevitable that a bee or two will get squished no matter how careful I am. It still makes me feel bad.

The queen is especially not something I ever want to kill, accidentally or otherwise. Only once have I considered killing one on purpose, and even contemplating it was nearly unbearable.

Here's the thing: My hives are in an area that is still relatively safe from the aggressive Africanized ("killer") bees. In the cool climate of the San Francisco area, where it only rarely gets above the low 70s, the hope is that it's too cold for the heat-loving bees to thrive and completely take over the local population as has been a more realistic fear for the hotter part of the state, only a few hundred miles south and east of here. However, just because they might not thrive here doesn't mean they can't live here long enough to spread some of their genes into local groups.

In early summer, one of my hives started giving me trouble. Bees from it began attacking me before I even got close, while I was working a hive on the far end of my bee yard, or even while I was hanging up clothes. To go in required a great deal of courage. I made the mistake of letting my most fearless apprentice help me work it one day. She had never been stung before, but this time, the bees noticed her black socks before I did and within a few seconds she got three quick stings on the ankle. (Never wear dark fuzzy anything when working bees—they think you're a bear or something.) She retreated, having lost her enthusiasm about bees, and I was left alone with the monsters as they buzzed angrily at the net that covered my face. These weren't just warning buzzes and butts; they were trying for my tender spots like the lips, nose, and eyes.

I made a quick inspection of the honey stores, was too concerned about my apprentice and my own well being to go deeper, and quickly closed up the hive again. At least a dozen angry bees followed me fifty feet to the back door, which is really unusual, and a few of them followed me inside, which is unheard of. That was bad. What was worse was hearing shrieks of horror an hour later—a few of the bees began aggressively buzzing a housemate as she emerged from her car, 100 feet and the opposite side of the house from the bee yard.

This wasn't good. This hive had gone from "having a bad day" grouchy to dangerous. Considering the various possible explanations, I decided the most likely was that perhaps a new queen had gotten its genetics mixed with drones from a hyperaggressive, maybe Africanized, hive. Whatever the case, things had gotten to a point that I'd have to find the queen and replace her with a gentler one. Part of the process of getting the aggressive bees to accept a new queen required the death of the hive's present queen. I didn't want to do it, but after consulting with other beekeepers, I didn't seem to have a choice. I felt like a cowboy having to shoot a favorite horse, but I felt like I had no choice, that sometimes a beekeeper's got to do what a beekeeper's got to do, etc.

Well, to make a long story short, after an elaborate, several-stage intervention to make it possible to find the queen without being stung to death, I discovered the real reason the hive was acting so aggressively. The huge hive didn't have a bad queen; it didn't have any queen at all. Apparently, the original queen had left with a swarm, and something had happened to the new queen that had hatched to replace her. (Quite probably she had been hit by a fast-moving truck or picked out of the air by a bird when she left the hive seeking to mate.) There was no queen, and no brood or eggs left to grow a new one, and the bees were in a surly, excitable, nihilistic mood. I borrowed a few frames with eggs and brood from my gentlest hive so they could grow a new queen, and the presence of brood and the hope of having a new queen immediately brought the angry anarchists back to their normal, peaceful selves.

Serious Design Flaw!

Complaint/Suggestion Ticket #338h3N0H

RE: Queen Succession Defect—IMMEDIATE ATTENTION!

Dear bee programmers,

Much of your work is absolutely first rate. Really! But there is one "bug" in your bugs that I really am surprised you haven't fixed by now, and I really hope you'll take care of in the upgrades. No, this isn't the sting problem, in which the guard bees automatically die when they sting. Yes, I still think that's a serious flaw. (What a waste. Do rattlesnakes die after only one bite?) And, yes, I am still underwhelmed by your response of "that's a hardware problem and we don't handle those," but this is even more serious, something that has doomed entire hives to oblivion when the predictable happens.

Here's the problem: the queen bee succession issue. As you know, here's the procedure when worker bees decide it's time to replace their queen and mother with one of their sisters. They grow several new queens in queen cups hidden around the hive. That's great, redundancy and all that. But what happens next? The first queen that emerges immediately goes about stinging the other queens to death while they're still trapped inside their queen cups. That's weird enough, but the other queens actually help her in the process, "piping" a high-pitched call that essentially means, "Here I am! Come and kill me!"

That's wasteful, even in the best of situations. I mean, why can't there be more than one queen? A hive with several queens would be much stronger. It's not as if they'd engage in power struggles, because they aren't really queens but are more like glorified egg layers. If the queens are sisters, why can't they just be programmed to get along and lay eggs side by side?

Okay, maybe that's beyond your programming abilities. How about at least reworking the sequence a little? Because, as you know, the queen kills off all of her rivals before she makes her mating flight. Now all of the hive's eggs are in one basket, both figuratively and literally. If the original queen has died or gone eggless, there's barely enough time to grow a new queen from one of her larvae, and if that doesn't work out, then there's a problem. A *big* problem, as in the survival of the entire hive is at stake.

Here's what really needs to be changed: Don't have her kill off her rivals so quickly. Don't have the long knives come out before she makes her mating flight. You know what can happen out there. She can get lost, get caught in a spider web, get picked off by a blue jay, or end up splattered on a windshield. If that happens, all those carefully programmed redundancies are for naught.

Even if she survives, there's a chance that she'll end up being infertile for some reason. That's why I make these recommendations, if you're really married to the "one queen per hive" model:

1. Ideally, there'd be understudy queens ready to spring into action if she turns out to be infertile. At the very least, put off the killing until she returns safely from her flight.

2. Why not have all of them take the flight, have a little fun with the boys, and come back to the hive? You could make it so that the first one gets past the guards, and the extra queens are barred at the door. That would eliminate the problem that your "sting 'em to death inside their cups" behavior was intended to fix. Having the queens engage in a fair fight adds a risk of no clear victor, with all of them wounded or killed.

3. Ideally, after they were barred at the hive door, they'd then go off and start their own hives, like some of the wild bee species in which the queen starts out alone building her nest, and then grows her court and workers.

I realize this might be beyond your skills and budget. Thanks, though, for considering my ideas.

Best, Jack Mingo

BETTER HOLMES & GARDENS

What did Sherlock Holmes do when he retired? Elementary, my dear Watson. He moved to Sussex and became a beekeeper, according to *His Last Bow*, Arthur Conan Doyle's last collection of Holmes stories:

"But you had retired, Holmes. We heard of you as living the life of a hermit among your bees and your books in a small farm upon the South Downs."

"Exactly, Watson. Here is the fruit of my leisured ease, the magnum opus of my latter years." He picked up the volume from the table and read out the whole title, *Practical Handbook of Bee Culture, with some Observations upon the Segregation of the Queen.* Alone I did it. Behold the fruit of pensive nights and laborious days, when I watched the little working gangs as once I watched the criminal world of London."

For many years, the bank located at 221B Baker Street in London employed a full-time secretary to answer letters from people asking Sherlock Holmes for help. Along with those letter writers, I only wish Holmes had been real and had indeed written that book. Bee behavior is still sometimes maddeningly mysterious; his keen skill at observation and deduction would have been very welcome indeed.

None of Your Beeswax

Bees, of course, use beeswax as the masonry of their architecture. It takes six to ten pounds of honey to make a pound of wax. The wax itself is fragile, but the hexagonal structure is a very light, very efficient, and very strong way to arrange their living and storage space.

Have you thought about where the wax comes from? It's sort of like bee dandruff, produced by glands under abdominal plates, that falls off in almost transparent little flakes. When heated up to about 91 degrees Fahrenheit and chewed by the bees, it becomes soft and whiter colored, allowing it to be molded into the familiar honeycomb shapes. The flakes are produced by young worker bees who stay hive-bound for the first weeks of their lives; when they get old enough to fly, the wax-producing glands become inactive.

Having too much honey and too little storage space stimulates the production of wax. Here's how it works: When the field bees unload their load of nectar into the honey storage "stomachs" of the inside bees, the nectar doesn't normally stay there very long, but gets downloaded into empty cells.

But what if there's not enough storage space? When the worker bee's stomach is full, it stimulates the bee's wax glands. They go into overdrive, dropping flakes all over the place. These flakes stimulate other bees to pick them up and start forming them into cell walls, which continues until there's enough storage space again.

Waxing Eloquently

Beeswax has, of course, been very useful to the bees, but it also has had a very long history as humanity's first plastic. In history, it has been used for sculpting metal, conditioning the strings for bows and crossbows, sealing and lubricating guns and musket balls, providing light, and even filling painful cavities in emergency dentistry.

Even now, although worldwide beeswax production isn't very much—only about 10,000 tons a year—beeswax is still widely used. You probably know about beeswax candles, and maybe about beeswax cosmetics and furniture polish, but did you know about these other uses for it?

- As a chewing gum.
- As a coating to preserve cheese as it ages.
- As polish for premium candies, including Jelly Bellies and Haribo gummi bears.
- In the molded French cakes called canelés, as a thin layer that gives a glossy, dark crust.
- Mixed with petroleum jelly, as the main ingredient in "bone wax," smeared by surgeons to control bleeding from bone surfaces during reconstructive surgery.

- As an ingredient in cosmetics, lip balms, mustache waxes, shoe polishes, and premium crayons.
- As a masking liquid for batik dyeing, protecting cloth that's to remain undyed.
- As the masking material when painting those elaborate Ukrainian Easter eggs.
- As a binder to hold together reeds on oboes, bassoons, and accordions.
- As the secret ingredient for a "thumb roll" on a tambourine, providing friction as you rub your thumb along it, sounding like a drum roll.
- As the etched coating on early "wax" phonograph records.
- Mixed with pine pitch and sawdust, as "cutler's resin," used by knife makers to attach blades firmly into handles.
- As a handyman's friend, fixing squeaks, freeing sticky drawers, and lubricating screws and locking them in place.
- As the sculpted form for the ancient "lost wax" casting process. It's coated with layers of clay and fired in a kiln. The wax burns away, leaving a mold into which bronze can be poured.
- As a sealing coat for cast metals (including, presumably, the sculptures above).

- As a whip coating. When some whip makers shifted from rawhide to cheaper, more durable nylon, customers complained that the weight wasn't quite right. It turned out that dipping the nylon in beeswax added the needed heft to make them just right.
- As a chicken plucker's plucking aid. Floating a layer of molten wax on top of boiling water sticks the feathers together so they come off together in a large clump instead of as a snowstorm of loose feathers.

Sharper Than a Serpent's Tooth

May 17, 2006

Flat on my back on a hard, cold sidewalk looking up at storefronts far from my house. It was not a place where I expected to suddenly wake up from a dead sleep. I rolled to the side and got up on one elbow. "Lie down! Lie down!" an unfamiliar voice said. What was going on? I could hear my wife's voice. I could see people in uniform and the open door of my car. It started coming back to me.

About fifteen minutes earlier, I had gotten stung a couple of times. Usually, the worst that would happen would be a grotesquely swelling limb, but this time had felt different. There was a metallic taste in my mouth, I felt light-headed, and my underarms and crotch were starting to itch, symptoms of a full body allergic reaction. I knew there was at least a slight chance it was developing into something serious. I'd had this experience a couple of times when allergy shots had gone bad, and it was scary when my throat began swelling shut, threatening to suffocate me like a python necktie.

I didn't want to panic my wife, but I didn't want to die either. The symptoms were still just a whisper. I figured it made sense to begin heading toward the hospital, only a few miles away, with the option of going in . . . or turning back if there were no problems.

"Erin, can you drive me to the hospital?"

I went upstairs, got my wallet, chewed a Benadryl tablet to speed up its absorption (it tasted terrible), and headed out to the car.

I had been returning some empty combs to my biggest hive without my bee suit (which has not been too unusual for small incursions into the hive) and had gotten stung very thoroughly by two conscientious guard bees. They stung deep and hard like professionals, not half-heartedly like some bees that were only putting in their time waiting for a shot at nectar collecting.

My fault, really. For reasons of ill-preparation, I had opened the colony twice already that afternoon, and apparently I had very much worn out my welcome. My hands being full of equipment, I couldn't immediately scrape off the disembodied bee stingers, so, fascinated, I had watched the now-beeless pieces of abdomen posthumously pumping toxins into my skin for a long twenty seconds. Feeling both otherworldly and agitated, I had made short work of what I was doing and retreated to the house, looking for both an antihistamine and my wife.

"I'll call 911," she had said.

"No." I was between insurance plans and did a quick calculation of an ambulance's cost per mile to the local hospital. A hundred dollars a mile? Maybe more?

"Can you drive? It's not really an emergency, just a precaution, and we'd be there before an ambulance got here." My voice sounded a little high and weird but not raspy, maybe just from nerves. No sign yet of a swelling, suffocating throat. I figured we'd make it. She reluctantly agreed to try.

As Erin rolled through the stop signs in our neighborhood, I had called 911 to ask them to let the local hospital know we were on our way, and why. They strongly suggested we pull off the road and let them send an ambulance, but we were already partway there, making good time, and the symptoms weren't appreciably worse so far.

That changed nearly as soon as I hung up. That's also when we hit the school traffic. Neither of us had anticipated that we'd be passing a high school at the end of the school day. Cars were jammed up and double-parked, and insolently sauntering students in small groups took their turn at making sure the crosswalks stayed blocked for minutes at a time.

I felt increasingly light-headed but still had no trouble breathing, so I passed the time morbidly reviewing with my wife the things to tell my daughter, baby grandson, and step-kids in case I didn't make it. I began having a weird sensation of the world around me getting weirdly brighter (guessing that, for some reason, my eyes were dilating). The brightness increased to the point where my field of vision slowly bleached to white. I remember noting to my wife with some detachment that it was exactly like the fade-to-white camera effect *Six Feet Under* used whenever someone died. I was both scared and fascinated (my wife, much more the former and much less the latter). I managed to report "that's weird, I can't see anything anymore" before whiting out completely.

I regained consciousness and eyesight long enough to realize that we'd made it about six blocks in what felt like a split second. I stayed awake long enough to compliment her on finessing through another jam-up before I peacefully detached from this world again.

Well, it was peaceful for me, anyway. For Erin, it was undoubtedly hell. As I slouched, blissfully unconscious, I was unaware as she screamed into the phone and 911 kept cutting off. Didn't hear her shouting and shaking me as I slumped forward in slack-jawed unconsciousness. Didn't feel her park the car or hear her plead with struck-dumb bystanders to help get me onto the sidewalk as the 911 operator had instructed.

Nor did I see Jesus, Buddha, my dad, my '65 Chevy convertible, or any other dead loved ones lolling around at the end of a tunnel.

When I did recover consciousness, I was aware that she was trying to drag me single-handedly out the passenger door. I tried to tell her that I could probably get myself out, but my efforts at communicating were not apparently as effective as I thought they'd be. Finally, she got two guys to stop watching and help lower me onto my back on the pavement.

Even before I had terra firmly under me, we could hear the sirens of the rescue workers. I was groggily conscious, and my eyesight was back again. A miracle! (Unfortunately, one of the things that kicks in last is good judgment of when humor is appropriate. As the rescue folks got me into the ambulance, a fire truck also pulled up with siren screaming. "A fire truck? Am I in danger of going up in flames?" I quipped. They didn't get the joke, and responded with "No, you're going to be all right," as if speaking to a delirious and somewhat slow child. Tough crowd.)

They hooked me up to machines, gave me oxygen, and perhaps some pseudoephedrine, stuck needles into me, and shouted out blood pressure numbers. By now I felt as conscious and as coherent as I ever was but glad to be lying down. I waved to Erin to assure her that she was still stuck with me and was pleased to see our neighbor Linda and her son Nicholas suddenly appear to embrace her (coincidentally they'd been driving by and saw the commotion). My breathing became fine, and my blood pressure quickly rose from weirdly low to more or less normal.

Eventually, we headed to the hospital, where I was forced to spend a few hours loafing around while they observed me and ran up my bill. My daughter Elana, who lived four blocks from the hospital, literally ran over with seven-month-old Acton, who spent the time inspecting oxygen tubes on my otherwise familiar face.

The drama had ended and changed to the tedium of waiting to be discharged. I went home with a Benadryl high, a prescription for EpiPens, and lots of things to mull over. The next time, I convinced a very reluctant Erin to do the bee interaction, but her resistance was matched by my eagerness to get back in the saddle.

Since then, I have upgraded my gloves to make it harder to sting me through them, but I still get three to six stings a year. I still keep my EpiPens nearby and review the instructions often but have never had to use them. I sometimes wonder if, upon feeling the beginning of a mild allergic reaction, I just panicked and fainted. I don't quite know how to feel about that.

Regardless, it's sobering to have a hobby that could conceivably kill me someday, but frankly also a little exciting. Sure, it's not skydiving or car racing, but every time I get ready to work the hives, I feel a tinge of danger to remind me that I'm not just a beekeeper, I'm an *extreme* beekeeper. I laugh at death, and "Danger" is my middle name.

Epilogue

Oh, Little Town of Buzz Buzz Buzz

Another year passes, another midnight on Christmas Eve, in what threatens to become an annual tradition. Again, the coldness of the evening, the cool, moist hives, my warm ear lowered gingerly to a cold, damp hive, listening to the solstice bee carols, the timeless melody that sounds like the purring of something, wild and eternal. . . . Sure beats listening for hooves, bells, and heavy steps on the rooftop.

About the Author

Jack Mingo is the author/co-author of more than 50 books, including *Doctors Killed George Washington* and *Random Kinds of Factness*. His work has appeared in hundreds of publications, including *The New York Times Sunday Magazine, Salon, Washington Post, Readers Digest, Wall Street Journal,* and *National Enquirer*. Mingo keeps half a million bees in six backyard hives in Alameda, an island community in the San Francisco Bay. Each year, his hard-working bees produce as many as 59 gallons of honey (650 pounds, or 472 pint jars), some of which he sells in select local establishments.

To Our Readers

Mango Publishing, established in 2014, publishes an eclectic list of books by diverse authors—both new and established voices—on topics ranging from business, personal growth, women's empowerment, LGBTQ studies, health, and spirituality to history, popular culture, time management, decluttering, lifestyle, mental wellness, aging, and sustainable living. We were recently named 2019 and 2020's #1 fastest growing independent publisher by Publishers Weekly. Our success is driven by our main goal, which is to publish high quality books that will entertain readers as well as make a positive difference in their lives.

Our readers are our most important resource; we value your input, suggestions, and ideas. We'd love to hear from you—after all, we are publishing books for you!

Please stay in touch with us and follow us at:

Facebook: Mango Publishing
Twitter: @MangoPublishing
Instagram: @MangoPublishing
LinkedIn: Mango Publishing
Pinterest: Mango Publishing
Newsletter: mangopublishinggroup.com/newsletter

Join us on Mango's journey to reinvent publishing, one book at a time.

Printed in the USA
CPSIA information can be obtained
at www.ICGtesting.com
JSHW031714140824
68134JS00038B/3687